Integrated Circuit
Design, Fabrication and Test

Integrated Circuit Design, Fabrication and Test

Peter Shepherd

School of Electronic and Electrical Engineering
University of Bath

McGraw-Hill

New York San Francisco Washington, D.C. Auckland Bogotá
Caracas Lisbon London Madrid Mexico City Milan
Montreal New Delhi San Juan Singapore
Sydney Tokyo Toronto

McGraw-Hill

A Division of The McGraw-Hill Companies

1 2 3 4 5 6 7 8 9 0 BKP/BKP 9 0 1 0 9 8 7 6

ISBN 0-07-057278-X

First published 1996 by MACMILLAN PRESS LTD
Houndmills, Basingstoke, Hampshire RG21 6XS
and London.

Printed and bound by Quebecor/Book Press.

This book is printed on recycled, acid-free paper containing a minimum of 50% recycled, de-inked fiber.

Contents

Preface

The world of the monolithic integrated circuit (the 'chip') had its beginnings in the early 1960s and, over the intervening years, has seen a massive increase in the complexity of circuit, the number of chips produced, the speed of the circuits and the different applications to which they are put. It can be truly said that there has been a revolution in electronic engineering due to the chip. Corresponding with the growth of the chip market has been the number of books available that describe the various aspects of chip design, technology, manufacture and, more recently, computer aids available to the engineer. The question you may justifiably ask as you hold this book in your hands is: why another?

The growth of available circuit complexity has remained on its exponential increase, with approximately a 100-fold increase in the number of available transistors on a chip every 10 years. This has taken the state of the art through from the initial small scale integration (SSI), about 10 devices per chip (1960), through medium scale integration (MSI), 1000 devices (1970), large scale integration (LSI), 10 000 devices (1980), to very large scale integration (VLSI), 1 million devices and beyond at the present day. However, until around 5 to 10 years ago, owing to the economics of chip design and manufacture, the only people involved in these activities were engineers employed by large electronics firms or in the IC foundries.

There has, however, been a second revolution within the chip technology area. With the more readily available computer aids, and IC foundries offering custom fabrication in multi-project wafers (MPWs) at affordable prices, the number of chip designers has increased out of all proportion to the general growth in the chip market. The field is now open to small companies to incorporate application-specific ICs (ASICs), which have been custom designed in-house, into their products. Research workers, academics and even undergraduate students are acquiring experience of the software tools, design techniques and manufacturing and testing processes. It is to this new breed of chip engineers that this book is directed.

The book is designed as an introduction to *all* the aspects of chip realization. As has already been noted, there are numerous books on the market that cover many of the aspects of IC engineering to varying degrees of depth. This book however discusses all aspects – design, manufacture and test of both digital and analogue circuits. It will help the engineer or student who is faced with the task of chip design for the first time and who requires some help through the mystifying maze of tools, technologies and techniques available. These aspects make the book unique.

As the book covers so many aspects and is designed at an affordable price, the depth of coverage is not great, but there is an extensive Bibliography at the end of each main chapter which the enthusiastic reader can use to pursue particular aspects to any required depth. These lists of further reading include books that the author has found to be of use in preparing the chapter material, and other relevant titles. References are given at the end of each chapter, if relevant.

Chapter 1 is an introduction to ICs, a summary of the design process and some of the considerations that must be made when a new IC is proposed. These include the choice of the appropriate technology and circuit architecture, planning the design, power considerations, testing and the economics of the product. The remaining chapters expand on these various aspects.

Chapter 2 describes the various families of technologies that are available for IC realization, detailing the properties of each and highlighting their various advantages and disadvantages.

Chapter 3 is involved in the detailed circuit design of the basic building blocks for each technology and covers both analogue and digital circuits.

In Chapter 4 the various architectures of the ICs are described, again detailing the advantages and disadvantages of each and when each should be used in preference to the others.

Chapter 5 is concerned with the computer tools that are available to the engineer in order to make the realization of an IC possible. The different function of each type of tool and the role it plays in the overall design process are detailed.

Chapter 6 describes the techniques for testing the devices, including the aspects of design for testability that must be incorporated into the design process from the start.

Appendix 1 covers the aspects of IC fabrication. Although it is not essential for the engineer to have a detailed knowledge of these processes to design an IC successfully (hence the positioning of this material outside the main text), these aspects do impinge on many of the design and test considerations. It is therefore strongly recommended to readers that if they have little or no knowledge of these aspects, then Appendix 1 material should be referred to before reading the main part of the book.

Appendix 2 describes an example of a simple IC design, covering virtually all of the aspects described in the main chapters. This IC is the product of a Final Year Design Study undertaken by students on the BEng Electronics and Communication Engineering Course at the University of Bath.

A small number of questions are included at the end of each chapter. The amount of mathematics used in the book has deliberately been kept to a minimum, but these exercises will give readers a chance to apply some of the ideas described in the text and to extend their insight into some of the concepts and problems.

I would like to thank John Martin of the University of Bath for his collaboration in the Final Year Design Project and for other useful conversations. Also Ian Walton and Richard Parkinson, the students who designed the example IC, for their permission to use the circuit description and results. I am grateful to European Silicon Structures for their permission to publish some of the details of the SOLO 1400 software and the IC fabrication route. I would also like to thank Mahmoud Al-Qutayri of DeMontfort University for permission to use some testing examples from his PhD thesis.

This book is dedicated to Barbara, who has made my life complete.

PETER SHEPHERD

Glossary of Abbreviations

a.c.	alternating current
ADC	Analogue–Digital Converter
ALSTTL	Advanced Low Power Schottky TTL
ALU	Arithmetic Logic Unit
ASIC	Application Specific Integrated Circuit
ASTTL	Advanced Schottky TTL
ATE	Automatic Test Equipment
ATPG	Automatic Test Pattern Generation
BILBO	Built-In Logic Block Observer
BIST	Built-In Self-Test
BJT	Bipolar Junction Transistor
BSDL	Boundary Scan Description Language
CAD	Computer-Aided Design
CAE	Computer-Aided Engineering
CAT	Computer-Aided Test
CIF	Caltech Intermediate Format
CMOS	Complementary Metal–Oxide Semiconductor
CMRR	Common Mode Rejection Ratio
CVD	Chemical Vapour Deposition
DAC	Digital–Analogue Converter
d.c.	direct current
DFT	Design For Testability
DIL	Dual-In-Line
DRAM	Dynamic RAM
DSP	Digital Signal Processing
DTL	Diode Transistor Logic
ECAD	Electronic Computer-Aided Design
ECL	Emitter Coupled Logic
EDIF	Electronic Design Interchange Format
EEPROM	Electronically Erasable Programmable ROM
EPROM	Erasable Programmable ROM

FET	Field Effect Transistor
FFT	Fast Fourier Transform
FPGA	Field Programmable Gate Array
FSM	Finite State Machine
GaAs	Gallium Arsenide
GOS	Gate Oxide Short
HBT	Heterojunction Bipolar Transistor
HDL	Hardware Description Language
IC	Integrated Circuit
IEEE	Institute of Electrical and Electronic Engineers
InP	Indium Phosphide
I/O	Input–Output
JFET	Junction Field Effect Transistor
JTAG	Joint Test Action Group
LFSR	Linear Feedback Shift Register
LPE	Liquid Phase Epitaxy
LSB	Least Significant Bit
LSI	Large Scale Integration
LSR	Linear Shift Register
LSSD	Level Sensitive Scan Design
LSTTL	Low Power Schottky TTL
MBE	Molecular Beam Epitaxy
MCM	Multi-Chip Module
MESFET	MEtal–Semiconductor FET
MOCVD	Metal Organic Chemical Vapour Deposition
MOS	Metal–Oxide–Semiconductor
MPW	Multi-Project Wafer
MSB	Most Significant Bit
MSI	Medium Scale Integration
nMOS	n-channel MOS
PAL	Programmable Array Logic
PCB	Printed Circuit Board
PLA	Programmable Logic Array
PLD	Programmable Logic Device
pMOS	p-channel MOS
PRBS	Pseudo Random Binary Sequence

PROM	Programmable ROM
PSRR	Power Supply Rejection Ratio
QTAG	Quality Test Action Group
RAM	Random Access Memory
r.f.	radio frequency
RIBE	Reactive Ion Beam Etching
RIE	Reactive Ion Etching
ROM	Read Only Memory
RTL	Resistor Transistor Logic (Chapter 2)
RTL	Register Transfer Level (Chapter 5)
s-c	switched-capacitor
SDI	Scan Data Input
SDO	Scan Data Output
SI	Semi-Insulating
SPICE	Simulation Program with Integrated Circuit Emphasis
SRAM	Static RAM
SRL	Shift Register Latch
SSI	Small Scale Integration
STTL	Schottky TTL
TC	Temperature Coefficient
TTL	Transistor Transistor Logic
ULM	Universal Logic Module
ULSI	Ultra Large Scale Integration
VHDL	VHSIC Hardware Description Language
VHSIC	Very High Speed Integrated Circuit
VLSI	Very Large Scale Integration
VPE	Vapour Phase Epitaxy
WSI	Wafer Scale Integration

1 The IC Design Process
Where do we start?

1.1 Introduction – a brief history of ICs

Integrated circuits (ICs) have their origin in the development of the solid-state equivalent of the thermionic valve – the transistor. Bipolar junction transistors (BJTs) were first developed in the late 1940s by Brattin, Bardeen and Schockley at Bell Laboratories, although point contact diodes ('cats whiskers') were in use before the Second World War, and the field effect transistor had been proposed but not successfully realized in the early 1930s.

Transistors continued to develop during the 1950s, originally based on germanium, and by the early 1960s cheap, reliable, silicon-based devices were commonly in use. These devices were all discrete, packaged individually, and then had to be mounted on circuit boards with other discrete components such as resistors and capacitors. The use of active devices in the realization of digital logic gates for computing processes had been employed for some time, based on valve circuits. The resulting computers were huge in size, and very inefficient, and the advantages of solid-state devices in terms of size and power consumption were soon to be appreciated. The integration of more than one component into a self-contained circuit was driven forward by this need. In fact, two forms of IC developed: the hybrid circuit, where the passive components and interconnections are manufactured using a 'thick-film' technique on a dielectric (usually alumina) substrate, the active devices being attached in their unpackaged 'chip' form to complete the circuit; and the alternative monolithic IC, where all the circuit components are generated in a 'thin-film' technique in which all the components and interconnections are on the semiconductor substrate. The latter has advantages in size and reliability and the hybrid IC is now very rare except for specialized applications, such as microwave integrated circuits.

So the first monolithic ICs, which emerged around 1960, consisted of just a few transistors, realizing individual logic gate functions or analogue amplifier circuits. Such circuits are termed small scale integration (SSI). As the process reliability and computer design tools developed, so the integration levels, in terms of the number of transistors per IC, have grown almost exponentially with time, as illustrated in Figure 1.1. With each decade of transistor numbers, there came a new term for the level of integration, as summarized below:

Nomenclature	*No. of transistors*
Small scale integration (SSI)	1–100
Medium scale integration (MSI)	100–1 000
Large scale integration (LSI)	1000–10 000
Very large scale integration (VLSI)	10 000–100 000
Ultra large scale integration (ULSI)	100 000–1000 000
Wafer scale integration (WSI)	over 1 000 000

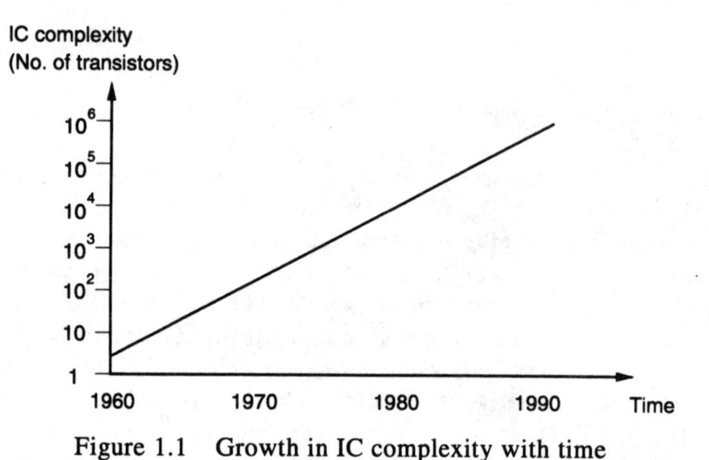

Figure 1.1 Growth in IC complexity with time

Note that technically the term VLSI is applied to circuits with transistor counts between ten thousand and a hundred thousand; but in practice the term has come into common use for virtually any monolithic IC, digital or analogue, and it is not unusual to see it applied to circuits that are MSI in complexity!

1.2 The design cycle

The design of a circuit comprising a million plus active devices is a daunting task. The complete design process from specification to IC realization has generally been beyond the scope of a single designer, as the total work load involved has often been of the order of several man-years. Teams of designers working together meant that new products could be completed within a year of real time. As competition for new markets and new products has accelerated in recent years, the complexity and power of the computer aids to design have reduced the work load to the order of man-months, and it is now within the scope of a single (very skilled!) designer to generate a complete IC within a reasonable time. However, to ensure quality of the product the design process must be well specified, with rigorous checks at each stage to ensure, as far as is practical, freedom from errors which would be expensive and time consuming to correct at a later stage of the process. This section briefly describes the various stages of the design cycle; many of these tasks will be developed in later chapters of the book.

The generalized design and production process for an IC is illustrated in the flowchart of Figure 1.2. The process starts with a specification for the IC. This

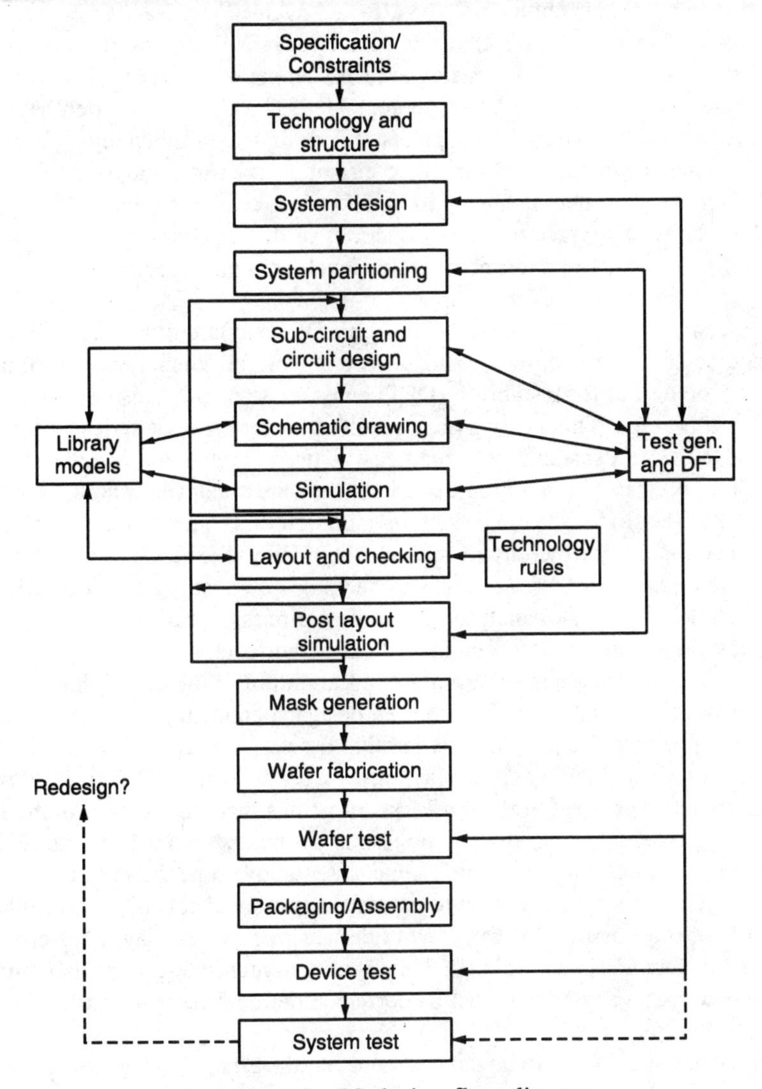

Figure 1.2 IC design flow diagram

will consist not only of the functionality specification, that is what the circuit does in terms of its primary inputs and outputs, but also such restraints as size, package type, power consumption, number of devices to be manufactured, expected product lifetime, etc. All of these considerations will affect the choice of technology, the overall design structure, testing considerations, etc., before the actual circuit design can begin.

Based on these various factors, as well as consideration of the economic factors, the type of circuit technology and the general structure of the chip are then determined. This choice may be affected by such constraints as the technologies and computer aids that are actually available to the designer and whether the circuit is digital, analogue or mixed-signal in nature. The overall system design and partitioning are then tackled. The circuit will often be broken down into sub-circuits which can be designed individually and then brought together to form the complete circuit. It is at this stage that testing must first be considered, since, in order to make the circuit sufficiently testable, the IC may have to include sections of dedicated testing circuitry. It is therefore essential to determine, at this level, how this circuitry will fit into the overall architecture of the IC.

The designer may now begin the design of the sub-circuit blocks, employing computer aids for drawing and library models of components or circuit blocks. The design for testability (DFT) considerations must also be borne in mind at this stage. This design stage results in a drawing, or series of drawings, of the circuit schematics, which can be used to generate circuit netlists, describing the parts in the circuit and how they are connected, these forming the input to the first stage of computer simulation. Such simulations check the basic functionality of the circuit without the time and expense of fabricating prototype circuits for test. As well as checking functionality the simulations may include 'worst-case' analyses to simulate process variations, measures of testability, performance as a function of temperature, etc.

The circuit may have to go around a redesign loop if the simulations show any performances that are not within the design specifications or constraints. However, once the designer is satisfied that the design is fully functional, the next stage to be tackled is circuit layout, which transfers the design from the schematic drawing level to the mask specifications that will be used in the IC fabrication process. Much of this process is automated under computer control, although there may be some human interaction from the designer. It is therefore vital that the layout patterns are thoroughly checked to ensure that none of the rules associated with layout dimensions and overlaying of different mask layers has been violated. This rigorous, intensive task is again done by computer, and any errors must be corrected before further progress can be made.

The next stage takes the layout data and regenerates the schematic circuit. There are two purposes here: firstly, it is an additional check that no errors have occurred in generating the mask data, provided the schematic is correctly reconstructed; secondly, additional data will be included in terms of the interconnections between the circuit elements. In the original schematic and its simulation, the simulator assumes that elements are connected with lines of zero length. In fact, the physical connections on the chip will provide additional delay and loading effects, and such effects may cause the circuit functionality to fall out of specification. If this is the case then some re-working of the layout will be required to shorten interconnections between critical elements.

Once the post-layout simulation is completely satisfactory, the layout data can be used to generate the physical masks which will then be used in the wafer fabrication process. Before the individual circuits on the wafer (die) are packaged, the circuit must be tested by probing the pads, in order to identify any circuits that are faulty owing to random fabrication defects. These are identified by placing an ink blob on them. The wafer is then diced up, and each individual die is mounted, bonded and packaged. A further device test is performed to check for any faults introduced in the packaging process. If the ICs are then to be mounted into a larger system before being released, this system will also go through a further testing stage.

The redesign route indicated in Figure 1.2 may occur if, for example, there are later reliability problems, the specifications are changed, or a new technology is to be used to generate the circuit.

1.3 Design considerations

As can be seen from the flow graph of Figure 1.2, there are a number of decisions and design considerations that must be made before the actual circuit design can be started. Often trade-offs will be made to choose the optimal design route, given the particular circuit specification and taking account of the economic factors (covered later in the chapter). Some of these considerations are discussed in more detail here.

1.3.1 Technology and architecture

There are several different technologies available for fabricating ICs. These are based on the different types of transistor that can be formed in the monolithic process. Each technology has particular advantages and disadvantages, and these are discussed in greater depth in Chapter 2. Broadly there are two forms of transistor, the bipolar junction transistor (BJT) and the metal–oxide–semiconductor field effect transistor (MOSFET). The latter comes in different forms: n-channel, p-channel, and enhancement and depletion modes (four combinations). MOSFET circuits may consist of only one channel type (such as nMOS) or complementary pairs of n-channel and p-channel devices (CMOS). Generally, bipolar circuits operate faster and with more power capability, nMOS provides good general-purpose digital circuits, and CMOS provides very low power consumption circuits. Digital circuits may be realized in any of the technologies, but analogue circuits only in bipolar and CMOS. Recently reliable mixed bipolar/CMOS circuits have become available (BiCMOS), so the advantages of each type can be utilized on a single chip. Mixed-signal analogue/digital circuits can also be generated using CMOS or BiCMOS.

The choice of technology will be influenced by a number of factors, primarily any particular constraints in the design specification, computer tools and routes to fabrication available to the designer, and cost. For example, for low-power specifications such as space-borne or hand-held equipment, the designer will almost certainly be constrained to a CMOS route. For extremely fast circuits, in the GHz range, the alternatives are specialized bipolar circuits or ICs fabricated on a gallium arsenide (GaAs) substrate and based on a metal–semiconductor Schottky gate FET structure (MESFET).

Having decided on the particular technology to be used, the choice of architecture of the IC will have to be determined. Again there are a number of alternatives, largely differing in the amount of scope that the designer has in customizing the internal structure of the IC. The range starts with programmable memory or logic array ICs, which are completely manufactured and packaged devices, but with micro-fuses built in that can be 'blown' by the designer to impose a pattern of customized logic on to the circuit. The next highest degree of design flexibility is based on the gate array device, in which the greater part of the IC structure is standardized, the designer having the capability of customizing the metal interconnections and so realizing the particular desired logic function. The custom chip is then packaged after the completion of fabrication. These first two approaches are only available for purely digital circuits – no analogue equivalent yet exists. The third approach is available for both digital and analogue ICs, and consists of a library of pre-designed components or 'standard cells' which the designer has freedom to place and connect up to achieve the particular circuit function(s) required.

All these approaches are examples of 'semi-custom' approaches, as the devices are pre-designed to a greater or lesser extent so that the designer is largely concerned only with the connection of internal components. They represent the cheapest (for small numbers of devices), quickest and easiest solution to an IC realization, and one or other of these approaches is likely to be the first design route that a new IC designer is likely to take. The learning curve is very short and the computer aids comprehensive – most of the design steps are fully automated and the designer does not need a detailed knowledge of the structure and operation of the components laid out on the circuit. However, these approaches are very unlikely to lead to optimal designs in terms of circuit size, speed, testability, etc. In order to achieve the best performance, a designer usually has to resort to a 'full-custom' approach which represents the other extreme of the range of architectures. Here the designer has complete freedom to specify the dimensions, positions and interconnection of the components, and it will be appreciated that the amount of work involved is therefore much greater than in the semi-custom approaches.

Although there is still a great deal of help available in the form of computer aids, the success of the design is critical on the designer's skill, knowledge and experience.

During the early years of ICs, all circuits were made using a full-custom approach and hence the number of IC designers was very few. It is only in recent years that computer tools and routes to fabrication, based on semi-custom approaches, have been made readily available to the electronic engineering community at large, leading to the total number of people involved in IC design mushrooming.

1.3.2 Top-down or bottom-up?

Having decided on the technology and architecture most suited for the particular circuit design under consideration, there is then the choice of the overall design strategy to be made. This may well be constrained by the software tools that the designer has chosen (or been constrained) to use. If these are a comprehensive package covering all the main design stages in an integrated form, they may well have checking operations specified at each stage and the designer must follow the prescribed design route.

Most designs will follow the lines of the flowchart of Figure 1.2. Here the design starts with a top-level specification and then breaks the problem down into smaller sections for separate design or consideration. Such an approach is termed 'top-down' and is considered by many to be optimal for good design practice and quality assurance.

There are alternative approaches to this scheme however. The other main approach is termed 'bottom-up' and starts from the lowest level of the circuit design, and usually this means the transistor level. From here the basic building blocks are designed – gates in the case of digital circuits; amplifiers, comparators, voltage references, etc. in the case of analogue. These can then be arranged into larger sub-circuits, circuits and systems. Mixtures of these two approaches can also be employed, working from both ends to meet in the middle, or starting from the middle and working out.

Each approach has various advantages and disadvantages. The top-down approach places constraints on the overall circuit or system from the start so that better control of the design can be exercised. However, it may lead to the definition of device parameters at a lower level of abstraction, which are difficult or impossible to realize practically. A reiteration in the design would then have to take place, seeking an alternative solution in terms of circuit structure or possibly relaxing design constraints.

The bottom-up approach ensures that each level provides a satisfactory solution to that stage's requirements, but as progress is made up the levels of hierarchy, there being no overall control of the end point, the final-level design may not be optimized in terms of size or cost.

Most semi-custom approaches are a mixture of top-down and bottom-up. Even pursuing the generalized top-down process of Figure 1.2, these approaches are based on the use of pre-designed circuits or architectures that have been formed in a bottom-up process.

1.3.3 Packaging and floorplanning

While it is not essential for the designer to have a precise knowledge of the details of the chip structure, particularly if automated layout tools are used, it is useful to be acquainted with the general structure of an IC and how it is packaged. This is illustrated in Figure 1.3, which shows a plan of an individual circuit die and how this is mounted in a typical device package. The actual semiconductor circuit area consists of two main parts. Firstly the central part, referred to as the core, contains the complete functional circuit as designed. The input and output signals, control and power supply lines must be taken from this central area to the outside world for connection to other cir-

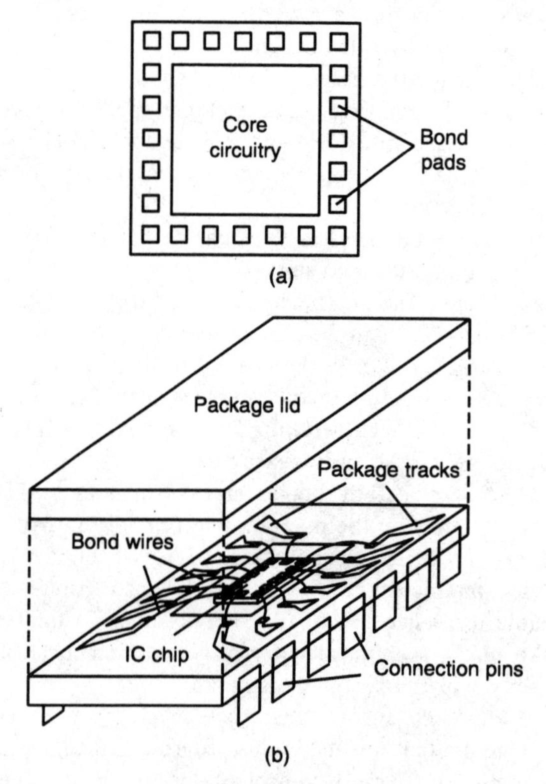

Figure 1.3 (a) Plan of bare IC; (b) assembly and packaging of IC

cuits and instruments. This is done through the use of bond pads which are placed around the periphery of the die, and these consist primarily of relatively large areas of metallization. When mounted in the package, the die is stuck on to a pedestal and connections are made between the bond pads on the die to corresponding pads on the package through thin metal (usually aluminium or gold) wires bonded to the pads by an ultrasonic or thermo-compression process. Metal tracks connect the package bond pads to the external pins forming the connection to the outside world. The packaging is completed by fixing a sealed lid on the device. The body of the package is usually either a plastic or a ceramic material.

For small to medium scale circuits, and most semi-custom circuits, the layout of the core circuitry is either fairly straightforward, under the control of automatic layout tools, or is constrained by the particular architecture of the circuit. For larger scale circuits, and especially when the IC is being designed by more than one engineer, much more care has to be taken in the layout of the core and the positioning of the input/output (IO) pads. If the design is a team effort then it is likely that different designers will tackle different sections of the circuit. Thus it is imperative that close control is exercized on the interconnections between these blocks and their relative size and position on the overall IC. This is done by a method termed floorplanning, where a large plan of the overall chip is created. The individual blocks of circuitry are represented by scale drawings, including information on the interconnections to the block, including signal, control and power lines. The floorplan originally consisted of a large paper drawing, the blocks being other pieces of paper that could be moved about to different places. Interconnections were drawn between the blocks and iterations were made in the design in terms of the position of the blocks to optimize the line routing, and often the blocks were redesigned to change their shape or the position of the connections to the block.

While the paper approach to floorplanning can still be taken, system software tools can now do the job more quickly and efficiently. Once the overall optimized IC design has been determined and all the individual blocks designed, the complete design can be merged to form the overall mask layouts.

1.3.4 Power considerations

All the circuit building blocks will require a power supply, although the number and voltage values will vary from technology to technology, and details are given in Chapter 2. What we are concerned with in this section are some of the global design implications.

The minimum, and the most usual, supply requirements are two lines, often a positive voltage and a ground connection, but certain analogue circuits may require a further negative supply in addition. However, since these power and ground lines have to be distributed around the IC, the pattern of metallization for these lines can become very complex. As all these supply lines are usually formed in the same metallisation layer it is very difficult, if not impossible, for the two lines to cross over, and this leads to a potentially difficult topological problem when laying out the circuitry. In the semi-custom approaches, for components such as gate arrays and standard cells, the task is relatively straightforward and the patterns are based on an interdigitated system, as illustrated in Figure 1.4.

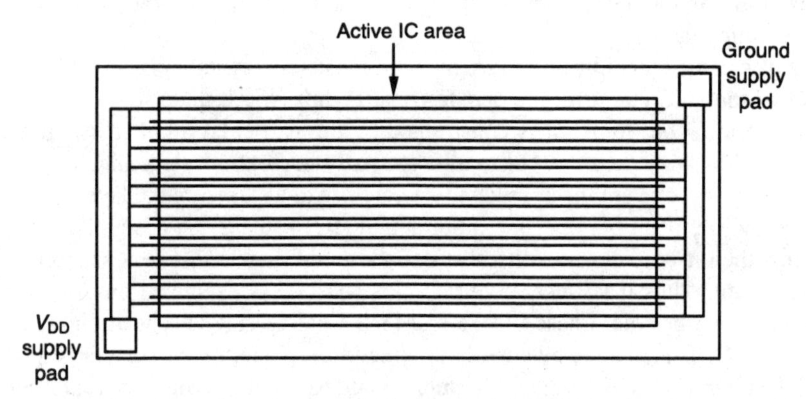

Figure 1.4 Interdigitated system of power and ground supplies in a semi-custom IC

The full-custom approach presents more of a problem, as blocks of circuitry can be placed anywhere within the core of the IC and great care has to be taken to ensure that the power supply lines can be routed successfully. The same problems can also arise with global signal lines, such as clocks, which must be distributed throughout the core. These restrictions must be considered at the floorplanning stage of the design so that the topology problem can be overcome. Often the positions of blocks, and the positions of connections to the blocks, will have to be redesigned to generate a successful overall routing plan.

The bonding pads, around the periphery of the IC, are often not just areas of metallization to which the bonding wires can be attached. They also provide the connections to the outside world, and the loads (usually capacitive in nature) attached to these external signals demand a relatively high current driving capability. This is in excess of the normal driving capability of the gates or output circuits of the core circuit components, and these signals therefore have to be 'buffered' using an additional active stage that does not change the nature of the signal except to provide a higher current driving capability. This is covered in more detail in Chapter 3 and all we must concern ourselves with here is to recognize that the output pads are active elements requiring a relatively large supply current to provide the necessary driving capability. The periphery circuit therefore requires its own power and ground lines and these may even be separate from the core supply lines, having their own pin connections.

Another consideration for the designer, apart from the topology of the supply lines, is the total current demands of the circuit being supplied. Some estimation of this must be made so that the power and ground lines are wide enough to supply the current without burning out. An associated design constraint is the resistance of the supply nets, which being global in nature can have a comparatively long length. The voltage drop along the supply lines should not be too great as to affect the overall operation of the circuit. Some design considerations are illustrated using Figure 1.5. Lines are usually rec-

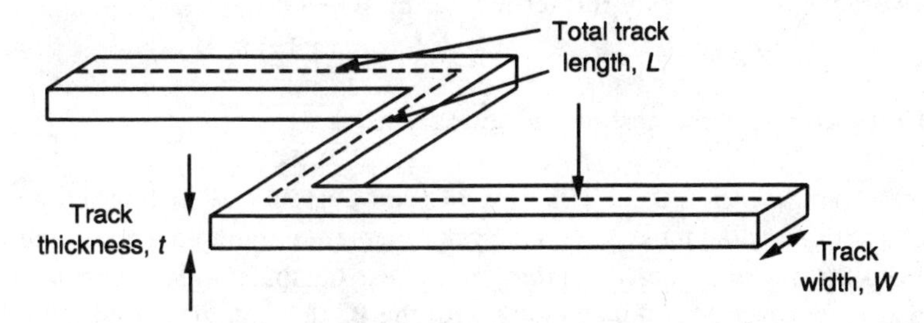

Figure 1.5 Dimension definitions for metal interconnection track

tangular in cross-section, with width W, thickness t and length L. The normal metal used is aluminium or perhaps one of its alloys. The critical parameters are the resistance of the line section and the current density within the line. The resistance of the line section is given by

$$R = \frac{\rho L}{A} = \frac{\rho L}{Wt}$$

where ρ is the resistivity of the material. The current density is given by

$$J = \frac{I}{A} = \frac{I}{Wt}$$

and the voltage drop along the line section is

$$V = IR$$

The design constraints are thus:

(1) A worst-case estimate must be made of the total current requirement of the circuitry being supplied by the line.
(2) To avoid problems of electromigration, in which there is movement of metal in high current densities and eventual breaking of the line, the current density must be set at a maximum limit. For aluminium this safe limit is approximately 10^9 A/m^2. Knowing I, this will set a minimum

value for the cross-sectional area. The thickness of the line is usually set by the fabrication process, so this sets a minimum value for W.

(3) An estimate is then made of the longest path of the supply track. From this, the resistance of the overall track can be calculated from the resistivity of the metallization.

(4) The voltage drop along the line at maximum current flow can then be calculated. A judgement must be made, based on the particular circuit technology, whether this drop in supply voltage will cause operational problems for the circuit. If so, the alternatives for redesign are to reduce the maximum length of line or, if this is not possible, to reduce the resistance of the line by further increasing its width.

1.4 Computer aids for design and manufacture

The massive growth in the availability of semi-custom and custom integrated circuits is due for the most part to the improvement of computer tools and the power of the machines on which they run. These computer aids to design, testing and layout were developed along with the IC technologies from the early 1960s, but it is only within the last decade or so that these tools have become available to any but the specialist IC designer. The detailed use of computer aided design (CAD) is covered in Chapter 5.

Up until around 20 years ago, most IC fabrication was based on standard components, such as the 7400 logic series. Any custom circuits that a manufacturer required to incorporate into a product or system, for example, had to be designed and fabricated by a third party. The customer would provide a detailed specification of the circuit operation and constraints, and the team of designers would then generate the circuit design, which would be manufactured often by yet another different company. The computing power required for such tasks as simulation and layout as well as the skills and experience of the design team usually forced all but the largest manufacturers to export the design stage to a specialist company. Likewise, only very large companies possessed their own fabrication facilities, the huge capital outlay making this facility uneconomical for 'occasional' use.

Since that time two factors have shifted the emphasis more towards 'in-house' IC design. The first has already been mentioned – the increase in computing power now available for a relatively small capital outlay. For example, the processing power and memory capabilities of a standard personal computer (PC) exceed that of a mainframe of 10 years ago for perhaps one-hundredth of the cost. This has made computer tools widely available. However, this is only half the story, as design skills still have to be developed for a custom IC design. This burden has been greatly eased by the development of semi-custom approaches such as programmable logic arrays, gate arrays and standard cell realizations. These approaches use largely standard circuits, in which large parts of the design have already been completed. The custom designer simply has to specify, for example, the logic specifications and the software will generate the circuit patterning for interconnections to complete the IC.

If the IC design has to be taken to the fabrication stage (which will be so in all cases except programmable logic circuits), there is still a problem in that it is very unlikely that the designer has access to in-house fabrication facilities. Often the number of these custom ICs to be fabricated is relatively small, and the high cost of mask fabrication does not make them economically viable. Realizing this problem, the fabrication companies (foundries) introduced a system by which several different IC designs could be incorporated on to one mask pattern, thus sharing the mask production cost between several customers. This again has only been possible because of computer tools that can take several different designs, in the form of computer data files, and merge them into a single mask pattern. This multi-project wafer (MPW) approach has made it possible for production of a few tens of any particular design to be economical. This provides an excellent route for prototyping ICs – often a semi-custom approach can be used to realize a prototype of a particular circuit. When the designer has checked this, a full-custom approach may be used to generate an optimized design, if the circuit is to be eventually manufactured in large numbers.

Readily available computing power, computer tools and MPW manufacturing have also meant that custom ICs have become available to the academic community, both for teaching and research purposes. Many of the software tools and fabrication facilities are made available at reduced prices. This 'loss-leader' approach is a common facility for academic institutions, the idea being that if students become familiar with a particular manufacturer's product at this stage, they will recommend or insist on these products when they move to industry.

In the UK in the mid 1980s, this access to IC realization was formalized with the establishment of the ECAD (Electronic Computer Aided Design) Initiative. Membership is open to all academic institutions on payment of an annual fee. This allows the organizing committee to negotiate contracts with software, hardware and fabrication suppliers that are more competitive than could be provided by a single institution. As a result, many tools were made available free of charge, many institutions have availed themselves of these facilities and the amount of IC design in academia has increased enormously. A few years afterwards, a similar initiative termed Eurochip was established across the countries of the European Community. The UK is part of this community, and so for a time there was an overlap of the two initiatives, with certain tools and facilities being available from both routes. This has recently been rationalized, so each scheme now has a different portfolio. Both operate a 'pay-as-you-use' scheme for the facilities rather than a fixed annual membership fee. Eurochip has recently been superseded by Europractise, who have reverted to an annual subscription. In the USA, a similar system exists,

MOSIS, which provides a prototyping service with fast turnaround and reduced costs, again making it ideal for use in academic institutions.

1.5 Testing

Details of testing are covered in Chapter 6. Here we look at some of the fundamental considerations that must be made at the outset of a design as regards verification of the product.

During the early development of IC technologies, design engineers designed the circuits and test engineers tested them. There was usually no input from the test engineer during the design stage – the designers would probably have resented this as an intrusion. The designer would often optimize a design purely from a performance point of view, with no regard as to how the circuit could be tested – that was the test engineer's problem. As the complexity of circuits grew, however, it soon became clear that circuits were becoming difficult, if not impossible, to test to a degree sufficient to satisfy the quality requirements of the customers or system designers who made use of the circuits. The concept of designing with testability in mind had to be adopted by the design engineer, and more interaction between the two camps was required. Design-for-testability (DFT) was born and it was recognized that testing could no longer be an afterthought, but a significant part of the overall design process.

The first developments in DFT were *ad-hoc* and consisted of better design practice, so that circuits could be more easily tested, and of circuit-specific adaptations to overcome particular difficulties. From these grew more formal approaches such as scan-path and built-in self-test. These DFT approaches carry a cost, however. Virtually all the techniques require extra pieces of circuitry which are dedicated to test and play no functional role in the normal circuit operation. The amount of extra circuitry is often termed 'overhead' and expressed as a percentage of the overall IC area. As the cost of IC fabrication is almost directly proportional to the substrate size, a trade-off must be made between this increase and the savings in testing cost. There is a law of diminishing returns in respect of this latter aspect, as illustrated in Figure 1.6. Here the individual costs of chip fabrication, testing and packaging are plotted as a function of the test circuit overhead. The test cost drops rapidly with the inclusion of a little DFT circuitry. As noted previously, the fabrication cost is a linear function, while assembly and packaging only increase very slightly, as a few extra external pins are required. It can be seen that the total cost passes through a minimum which is typically of the order of 10 per cent overhead. Military or safety-critical devices may have a higher testing specification, and so the higher cost will be borne.

The testing decisions to be made at the outset of the design will depend on the size of the circuit and the technology being used. Circuits realized by pro-

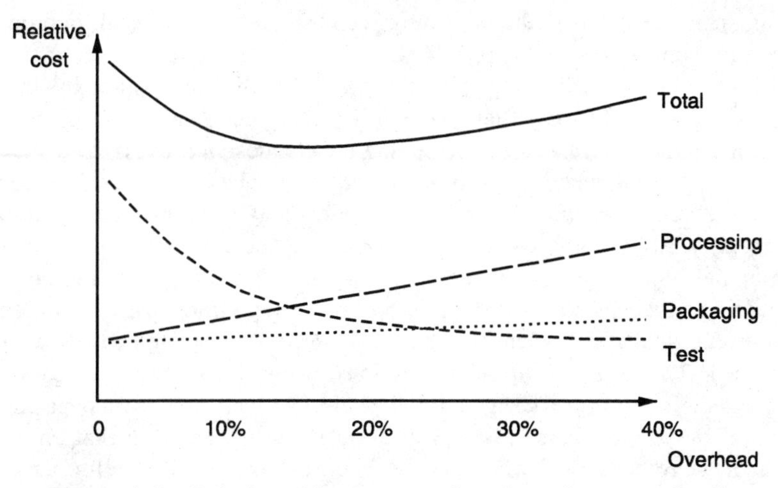

Figure 1.6 Relative chip cost as a function of test overhead

grammable logic or gate array approaches are unlikely to permit testing circuitry to be incorporated, as they are based on a fixed circuit architecture. These circuits are only of small to medium scale size, which should be sufficiently testable using good design practice and the formal test pattern generation approach. Semi-custom foundries often require the designer to provide a set of test input patterns with a minimum fault coverage which the foundries can use to test the circuits before assembly. This is sometimes a requirement before the design will be accepted for fabrication.

Circuits of LSI complexity and above will almost certainly require some formal DFT to be included. Again it depends on the actual size and circuit functions which are the most appropriate techniques to apply. When the design has been broken down into the various sub-circuits for separate design, testability analyses can be run on these individual blocks to identify potential problem areas and concentrate the testing circuitry where it is most required. This is included in the design flow graph of Figure 1.2 and will also have consequences on the floorplanning operations, as inclusions of such circuits as scan paths will alter the overall topology and layout of the IC.

1.6 Economics

Virtually every IC designed and fabricated, particularly for commercial purposes, will have some form of budgetary constraint, and this will influence most aspects of the design decisions. There may be other constraints in terms of which software tools and fabrication facilities are available, which will eliminate some options, but within the available choice the final route will almost certainly be guided by economics. We have already seen some of the design considerations in the preceding sections, and here we shall outline the overall picture.

There is no one best route to take for all IC realizations, since the choice depends on many factors, often with conflicting effects, and the designer must carefully balance all these before setting out the design process. The specification that accompanies the proposed circuit should be comprehensive enough to base the decision on; if not, the final product may be unrealizable or uneconomical, wasting much money and effort. The specification must of course define the function of the circuit, including details of the inputs and outputs. In addition the speed, or bandwidth, power consumption, size, weight, environmental factors and where the product is to be incorporated should also be provided. This is likely to put restrictions on the type of technology to be employed, for example CMOS, bipolar, etc., as well as the final packaged form of the device. It may be that the specifications are such that only one technology route is available to the designer. If this is the case a further analysis should be made on the degree of customization to be incorporated and on whether the final product is still viable, before committing to the product development.

Other factors now come into play. It is likely that the end user of the IC will have specified a maximum price that can be allowed for the completed product. The price per IC will depend not only on the cost of fabricating and testing each individual circuit, but there is a capital outlay in terms of design time, mask making, etc. which must be spread over the total number of ICs manufactured. So the designer must know from the outset the likely number of circuits to be manufactured.

If the total number is relatively small, say less than a few thousand, the full-custom approach will not be economical, because this approach requires the largest design effort and hence cost. To spread this cost over a small number of products would result in an excessively large add-on cost to each device. Full-custom approaches are only applicable when very large numbers of devices (typically at least 100 000) are to be manufactured. In this way the add-on cost due to the greater design effort can be offset by the facts that the design is likely to be better optimized, the circuit size smaller and hence each individual chip cheaper to produce.

For small numbers of product, the design must make use of semi-custom approaches such as gate array or standard cell, possibly through the MPW fabrication route. These approaches can be highly automated, with design times of, typically, a few weeks, or even less using silicon compilation techniques. In using MPW the cost of mask production is shared among several designs, so overhead costs are minimized. Although the circuit may not be optimized in terms of circuit size, speed or power consumption, the overall cost per chip is minimized. Provided that the original circuit specifications are not violated, this must be the chosen route.

If the device is to be developed as a commercial product, as opposed to a research or development aid, it is not sufficient simply to fit the cost within a particular budget, since the product should also return a certain profit level over its expected lifetime. The precise level is likely to depend on company decisions that are outside the influence of the designer, but he or she should be aware of the expected return.

The costs associated with semi-custom approaches tend to be largely fixed – fabrication costs are quoted on a per unit area basis, assembly costs depend on the particular package size and type, but again are fixed quotes. Some foundries incorporate circuit test, only supplying circuits that have passed the specified test. If the designer wishes to do extra testing, this will be an additional cost which will have to be estimated. The only other factors to be determined before the final cost per device can be calculated are the design time and the likely total number of devices to be manufactured.

With the full-custom approach there may be other factors involved, as more scope could be available for specifying the fabrication processes. In particular, the size of the wafer can be important. As the technology improves, so larger diameter wafers can be processed. When ICs were first fabricated the standard was a diameter of 1 inch. This soon increased to 2 inches and has since expanded further so that the standard nowadays is a 5 or 6 inch diameter. The larger the wafer, the more circuits can be manufactured in one process run. However, until a new technology becomes properly established there is a problem with yield. Yield is defined as the number of good circuits produced in a process run as a percentage of the total potential circuits. New technologies have a lower yield both in terms of a higher density of defects on each wafer, and wafer breakage, as larger circuits tend to be more fragile than smaller ones. With each change in wafer size there is an interim period when the two technologies are comparable in overall IC fabrication cost. This will remain the case until the yield of the newer technology improves to a degree where the extra ICs per wafer make a significant reduction in individual IC cost, resulting in the older technology becoming obsolete.

1.6.1 Example of an economic forecast

A company is to produce a new IC and a chip size of 20 mm^2 has been estimated, based on using a full-custom CMOS technology on a 6 inch wafer. Additionally, the process has a 95 per cent yield at the wafer fabrication stage and a 98 per cent yield at assembly. The design is to take 3 months to prepare and the first devices will be ready for delivery 3 months later. The product is expected to have a life of 5 years and it is anticipated that sales of 100 000 devices per year will be made, although to remain competitive the price will have to be reduced by 10 per cent per year. A 200 per cent return on capital outlay is expected from the product over its life. A forecast of cashflow and estimate of initial price have to be made.

The capital outlay consists of many factors, primarily the design team's salary plus overheads (taxes, fringe benefits, infrastructure items such as heating, lighting, space, etc.). In addition there are costs in technician support, computer usage and software licensing, equipment wear, documentation, test development and mask masking. Such an outlay may typically be of the order of £200,000 for a design of this size. This outlay will occur during the first 3 months when the design is being prepared. After the devices go into production, no sales will be possible for a further 3 months, and typically it will be a further 3 months before the accounts are settled and money starts to be recouped. The sales over the following five years must result in an overall profit of £400,000, that is an income of £600,000 plus the cost of producing the devices.

Typically a 6 inch (150 mm) wafer costs about £200 to process, including purchase of blank wafers, processing, probing and die separation. A 6 inch diameter wafer has an area of 17 671 mm^2 so about 880 die could be produced. Taking yield into account, this would result in $880 \times 0.95 \times 0.98 = 819$ good devices. Each device therefore costs around 25 pence to process. The package will probably cost a similar amount (depending on type and number of pins required). In addition, testing costs are typically 30 per cent of the production cost, resulting in a total device production cost of 65 pence.

We shall assume that these costs remain constant over the five year production cycle. At a rate of production of 100 000 per year, this results in an outlay of £65,000 per year and a total production cost of £325,000. So the product must return a grand total of £925,000. Suppose the initial price per unit is £x. This price has to be reduced by 10 per cent in each successive year, so the income generated by the sale of 100 000 devices per year is:

Year 1	100 000x
Year 2	90 000x
Year 3	81 000x
Year 4	72 900x
Year 5	65 610x
Total	409 510x

This results in an initial sale price of $x = 925\,000/409\,510 =$ £2.26. The overall cashflow is illustrated in Figure 1.7. This predicts a 'break-even' point soon after two years. The figures used above will of course depend on the precise product, but they are typical and illustrate some of the considerations that must be made in assessing a product's viability.

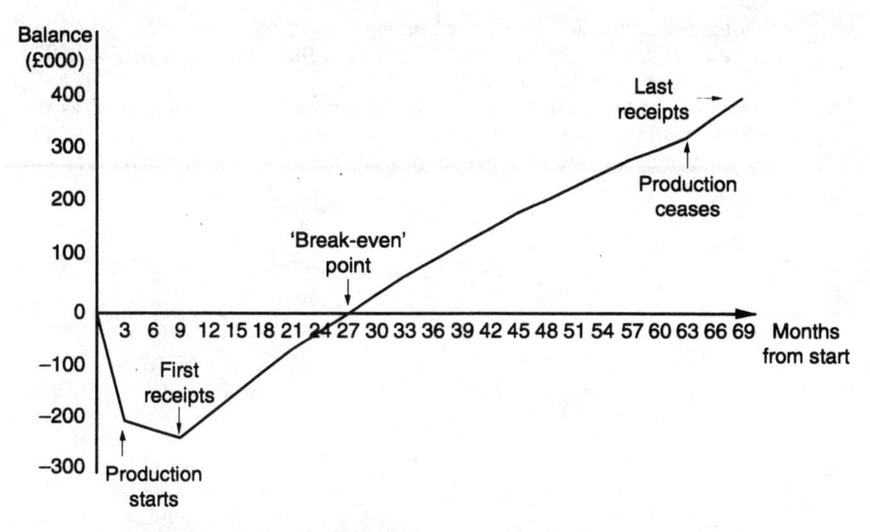

Figure 1.7 Cashflow forecast for design example

Bibliography

T.E. Dillinger, *VLSI Engineering*, Prentice-Hall, Englewood Cliffs, New Jersey, 1988.

D.A. Fraser, *The Physics of Semiconductor Devices*, 4th edn, Oxford University Press, Oxford, 1986.

R.L. Geiger, P.E. Allen and N.R. Strader, *VLSI Design Techniques for Analog and Digital Circuits*, McGraw-Hill, New York, 1990.

J. Mavor, M.A. Jack and P.B. Denyer, *Introduction to MOS LSI Design*, Addison-Wesley, London, 1983.

W. Wolf, *Modern VLSI Design – A CAD Based Approach*, Prentice-Hall International, Englewood Cliffs, New Jersey, 1993.

Questions

1.1. The flow diagram of Figure 1.2 represents a 'top-down' approach to design. How would the flowchart be altered if the design process includes a partial 'bottom-up' approach whereby standard cells are created for use in the circuit design phase?

1.2. An aluminium metallization is 5 μm in thickness, the effective total length is 10 mm and the highest current it is expected to conduct is 50 mA. Using the data in section 1.3.4, calculate the minimum line width if a 20 per cent safety margin is to be built into the current density. [12 μm]

1.3. For the line in question 1.2, calculate the resistance of the line given that the conductivity of the metal is 3×10^7 $\mathrm{S\,m^{-1}}$. Calculate the voltage drop at maximum current flow. Comment on the results. [5.6 Ω, 0.28 V]

1.4. Consider the example in section 1.6.1. An alternative technology based on 8 inch wafers becomes available. The process has only 90 per cent production yield and extra tooling, training and equipment add a further £100,000 to the set-up costs. In addition, the 8 inch wafer costs £300 to process. If all other parameters, costs (including packaging and testing) and timings remain the same, what would the initial price of the devices have to be to return the same profit of £400,000? Which technology would you recommend? [£2.47]

1.5. For the new technology in question 1.4, if the initial price was the same as the original technology, sketch the predicted cashflow–time graph. What are the 'break-even' time and the final overall profit? [36 months, £315,500]

2 IC Families
What technologies can we use?

2.1 Introduction

Although there is a vast array of discrete electronic components, both active and passive, available to the circuit designer, when circuits are of a monolithic, integrated form, most of the circuit is made up of transistors. A very few diodes, resistors and capacitors may be used, but components such as inductors, transformers and the thyristor family of components are virtually never realized monolithically.

Transistors can be used to realize most circuit functions, both analogue and digital. They can be used, for example, as load resistors, and occupy much less space than monolithic resistors and capacitors. As has been noted, area is a very important factor in the cost of producing an IC, so any reduction in area without compromising performance is to be welcomed.

Hence most circuit design is based on transistors, and these techniques are examined in Chapter 3. Here we shall look at the various generic families of circuits that are available, and the various advantages and disadvantages of each.

Transistors are three-terminal devices, most often configured as two-port components with one of the terminals common to the input and the output. In general, the output current or voltage is controlled by the input current or voltage. Transistors can be operated in either a linear mode, whereby the output parameter varies linearly with the input parameter, or in a non-linear mode whereby the output parameter is switched between its two extremes. The former is used largely in analogue circuits, whereas the latter provides the basis of digital operation.

There are two main forms of transistor, based on the way the device is fabricated. Although these two forms are structurally very different, the overall circuit behaviour of the two ports is very similar. The first is formed by three regions of alternately doped p- and n-type semiconductor and is termed a bipolar junction transistor (BJT). The second is formed by having a semiconductor channel, the resistance of which is altered by the voltage on a capacitive structure along the side of the channel. This type of device is termed a field effect transistor (FET). We shall now look in more detail at these two structures.

2.2 Transistor types

2.2.1 Bipolar junction transistor

The BJT consists of three separate regions of semiconductor, alternately doped p-type and n-type. Consequently there are two alternative forms: npn and pnp. Each region has a connection, termed emitter, base and collector, as shown in Figure 2.1, together with the standard component drawing. In the latter, the direction of current flow is indicated by an arrow.

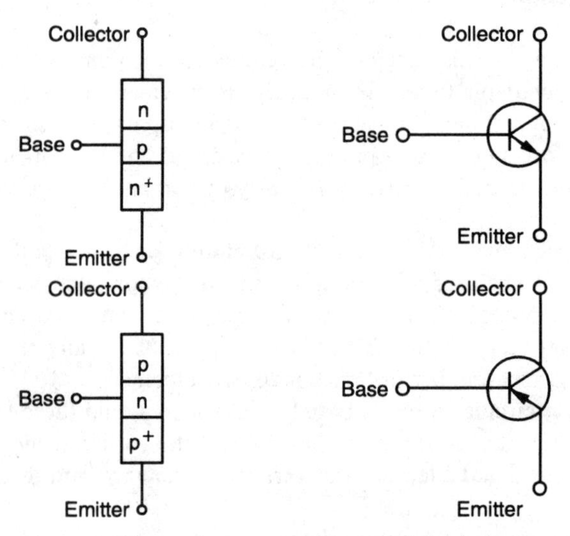

Figure 2.1 npn and pnp bipolar transistors

The device can be considered as two back-to-back p–n junctions. In the normal bias condition, the emitter–base junction is forward biased and the base–collector junction is reverse biased. In addition, the emitter is doped much more heavily than the base (as indicated by the + signs in Figure 2.1). This means that with the forward-biased emitter–base junction, majority carriers are injected from the emitter to the base (these are electrons in the npn device and holes in the pnp device). These majority carriers become minority carriers in the base, so recombination takes place. To minimize this effect the base is made very short, so most of the carriers reach the second junction where they are swept across under the influence of the reverse bias of the base–collector junction. Hence we have a main current flow from emitter through to collector, with a small current flow into the base region to make up for the recombination of carriers. Small variations in this base current result in large changes in the collector current, so the device acts as a current amplifier. This describes the linear operation of the device.

In addition, if the emitter–base junction is not forward biased sufficiently, little current injection occurs and, because of the reverse-biased base–collector junction, the resistance between emitter and collector is very high and the device is said to be cut-off. Alternatively, if a very high base current is injected, then variations in this current cannot be mirrored in variations in the collector current, the resistance between emitter and collector is very low and the device is said to be saturated. These two 'on–off' states are used in digital circuits, with the device being switched rapidly between the two. Such characteristics are summarized in a typical plot of the output characteristics (collector current against collector–emitter voltage as a function of base current) for an npn device, as shown in Figure 2.2. The characteristics for the pnp are very similar in form, but the polarity of the voltages and currents is reversed.

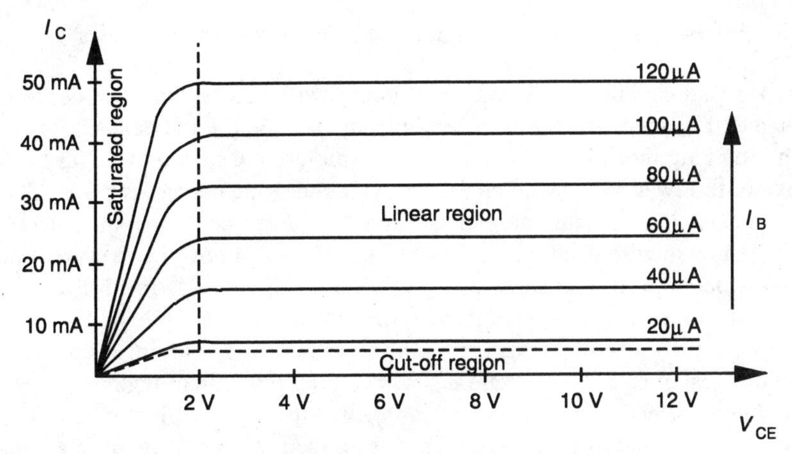

Figure 2.2 Output characteristics of bipolar junction transistor showing typical values of voltage and current

2.2.2 Field effect transistors

Field effect transistors (FETs) are based on the principle of altering the resistance of a section of semiconductor by application of a transverse field along the length of the device. The contacts at the two ends of the device, through which the main current flow passes, are termed the source and the drain. The contact by which the field controlling voltage is applied is termed the gate.

There are two basic forms of FET, the junction FET (JFET) and the metal–oxide–semiconductor FET (MOSFET). The JFET is primarily only seen in the form of discrete transistors, and only the MOSFET is monolithically integrated, so we will concentrate on this latter type. The basic MOSFET is very simple in structure, and is shown in Figure 2.3. It consists of a section

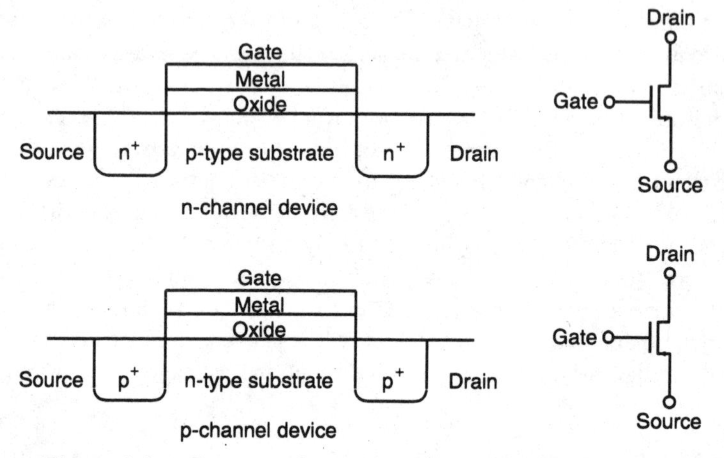

Figure 2.3 Cross-section and symbols of MOSFET devices

of doped semiconductor (termed the channel), with source and drain contacts at each end formed with wells of semiconductor doped in the opposite sense. Along the length of the channel there is a structure consisting of a thin layer of oxide, followed by a metal contact to which the gate contact is made. This forms a capacitor structure along the length of the device. In certain processes, the gate structure is not made of metal but a layer of doped polycrystalline semiconductor. Both n-channel and p-channel FETs are illustrated in Figure 2.3, along with the standard component drawings.

Consider the n-channel structure as shown in Figure 2.3, where there are no bias voltages applied. The junctions between the source/drain regions and the substrate are effectively back-to-back p–n junctions, so no current would flow between source and drain if a voltage was applied, as one of these junctions would always be reverse biased. (This differs from the BJT, which also consists of back-to-back p–n junctions, as the 'base' region here is very long and all the injected minority carriers would recombine before reaching the second junction.) We need to establish a conducting channel between source and drain, and this is done by applying a positive voltage to the gate. If this occurs gradually, it is found that first of all a depletion region is established beneath the gate structure, owing to the positively charged holes being repelled by the positive charge on the gate. As the gate voltage is increased, the negatively charged minority electrons are drawn up from the substrate towards the gate, forming an 'inversion' layer of n-type semiconductor and subsequently a conducting channel between the source and the drain. The gate voltage required to establish this conducting channel is termed the threshold voltage, V_T, and is a very important parameter in the design of FET circuits.

Once the channel is established, a current can be made to flow between source and drain by applying a positive voltage to the drain. The device then acts a resistor, with an almost linear relationship between the drain voltage and current flowing through the channel. This drain voltage is dropped along the length of the device, so the effective voltage between the gate and channel also varies, together with the thickness of the conducting channel, as shown in Figure 2.4. If the drain voltage is increased further, eventually the thickness of the channel at the drain end shrinks to almost zero. Then there is no significant increase in the current with increasing drain voltage, and the device is said to be 'pinched off'.

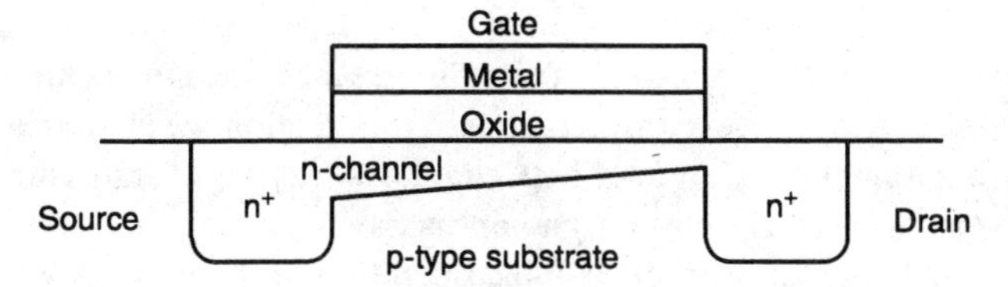

Figure 2.4 n-Channel MOSFET showing active channel

If the gate voltage is increased then a thicker channel is established with a lower resistance, so more current flow is allowed for a given drain voltage, and the device pinches off at a higher voltage. This leads to a set of output characteristics as shown in Figure 2.5. It can be seen that these are very similar in form to those of the BJT and so, although the physical structure of the two devices is very different, their behaviour as a circuit component is similar and they can be used for similar applications.

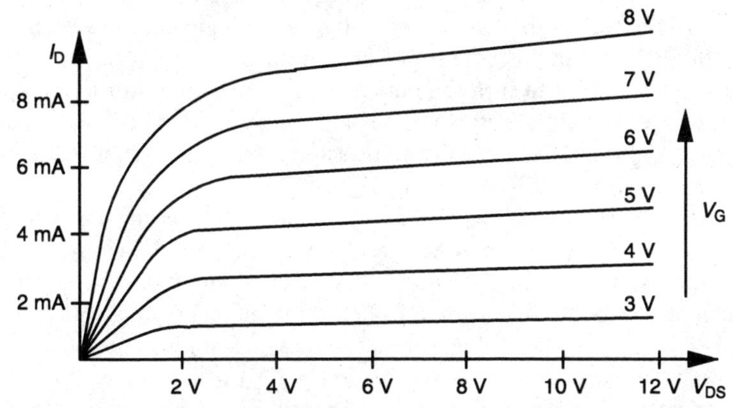

Figure 2.5 Output characteristics of MOSFET showing typical values of voltage and current

This basic form of FET is termed an enhancement mode device, as no current can flow without a gate bias voltage, so the behaviour is 'enhanced' by the gate bias. However, if a thin conducting channel is introduced below the gate structure by doping or ion implantation, then a drain current could flow without the need to bias the gate. Such devices are termed depletion mode devices since the behaviour can be 'depleted' by applying a negative gate bias. The behaviour can be further 'enhanced' by applying a positive gate bias.

Enhancement and depletion mode devices also exist for p-channel devices, but as with the difference between npn and pnp BJTs, the doping conditions, bias voltages and direction of current flows are all reversed.

2.2.3 Comparison of BJT and FET devices

The details of the different circuit technologies and design techniques are explored in the following two chapters, but in this section we shall take a brief look at the fundamental differences between the two types of transistor, which forms the basis of the different design approaches.

The main fundamental difference in the operation of the two devices, which reflects particularly in the modelling of their operations, is that the BJT is a current-controlled device, whereas the FET is a voltage-controlled device. This is reflected in the plotting of the output characteristics, Figures 2.2 and 2.5, where for the BJT the family of curves is shown as a function of the base current, I_B, whereas the FET characteristics are shown as a function of gate voltage V_G. In modelling terms this means that the BJT is represented by a current-controlled current source, with an associated current gain β, and the FET as a voltage-controlled current source with an associated transconductance, g_m.

These models are used primarily for linear, analogue operation. The digital modelling is based on the two extreme non-linear regions of the characteristics. In the BJT, the important parameters are the base–emitter voltage, or the base current, required to turn the device on, to define the threshold between the cut-off region and the active region, and the value of the collector–emitter voltage, V_{CEsat}, when the device is in the saturation region. For a silicon bipolar device, V_{BE} is typically 0.7 V and V_{CEsat} 0.2 V.

For the FET, the important parameter to define device turn-on is the gate threshold voltage, V_T, which varies greatly depending on the particular fabrication process, but typically may be 1–2 V. As the output curves for the FET are much more smooth than for the BJT, there is no real equivalent of the low value V_{CEsat}. The output voltage in a simple MOS digital circuit when the device is turned on can be relatively large ($\approx$ 1 V) which means that a lot of quiescent power is dissipated, and the corresponding logic threshold may be poorly defined. The consequences of this are explored in later sections.

In general, BJTs are likely to be used in higher supply voltage circuits. This is a consequence of the relatively large V_{BE} threshold value, which is largely a function of the semiconductor material rather than device geometry or doping, and so cannot be significantly altered. FET circuits can be operated at much lower supply voltages, with the resulting potential for much lower power operation.

Having discussed the basic devices, we shall now look at some of the subsequent circuit configurations that can be employed.

2.3 Digital circuits

Historically, when transistors superseded valves as the common active device in electronic circuits, the BJT was the only practical semiconductor device; FET circuits arrived soon after. For digital circuits, different families developed, based on the two types of device. For the BJT, two major groups exist: transistor–transistor logic (TTL) and emitter-coupled logic (ECL). For the FET, the two groups consist of simple MOS circuits, and a group made up of pairs of n-channel and p-channel devices, termed complementary MOS or CMOS. We shall look in detail at each of these implementations and compare their various advantages and disadvantages.

2.3.1 TTL

The basic BJT gate is based on the operation of the transistor in saturated and cut-off modes as the two logic states. Figure 2.6 shows the basic inverter circuit and its transfer characteristics. This formed the core of the earliest BJT logic; for example, by paralleling up devices, a NOR gate can be formed, as shown in Figure 2.7. These first gates were formed, as shown in the figures, by resistors and transistors, and this is termed resistor–transistor logic (RTL). Although, in theory, the output voltage from an RTL gate is approximately equal to V_{CC} for the logic 1 condition, if this then drives the input of a subsequent gate, the base current drawn into this second stage means that the out-

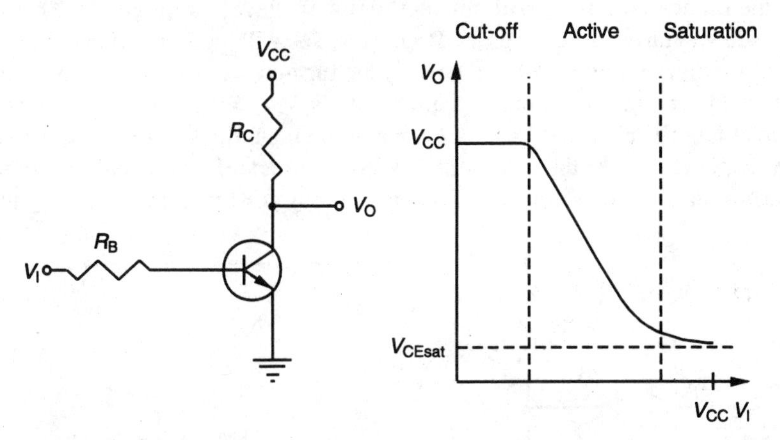

Figure 2.6 Basic bipolar inverter and transfer curve

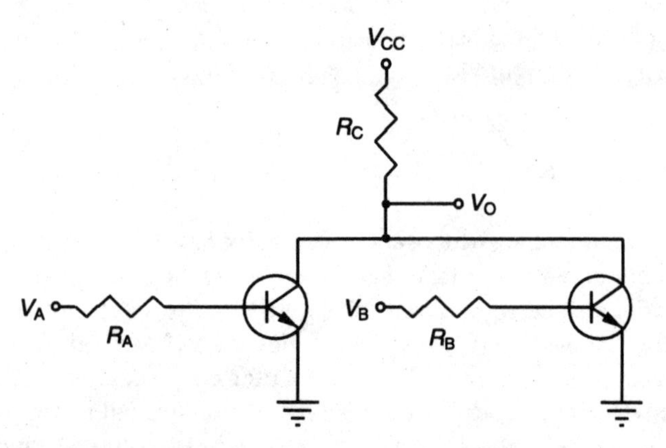

Figure 2.7 RTL NOR gate implementation

put voltage drops considerably, so the fan-out (number of gates that the output can drive) is limited. Additionally, the power consumption of RTL is fairly high, so this simple configuration was superseded.

The next stage in development of BJT logic circuits involved the use of diodes in combination with the transistors, and so is termed diode–transistor logic (DTL). An example of a NAND gate is shown in Figure 2.8. Qualitatively, if any or all of the inputs are held low ($\approx$ 0 V), then at least one of the diodes D_A to D_C will conduct, drawing current through R_A. This will pull down the voltage at point P and D_1, D_2 will be non-conducting. The base–emitter voltage will be below the turn-on voltage, the transistor is switched off and the output is approximately V_{CC}. When all three inputs are held at logic 1 (close to V_{CC}), then none of the input diodes conducts, the voltage at point P rises, the diodes D_1, D_2 conduct, and the base–emitter voltage rises to turn the transistor on and so the output voltage drops to V_{CEsat} (logic

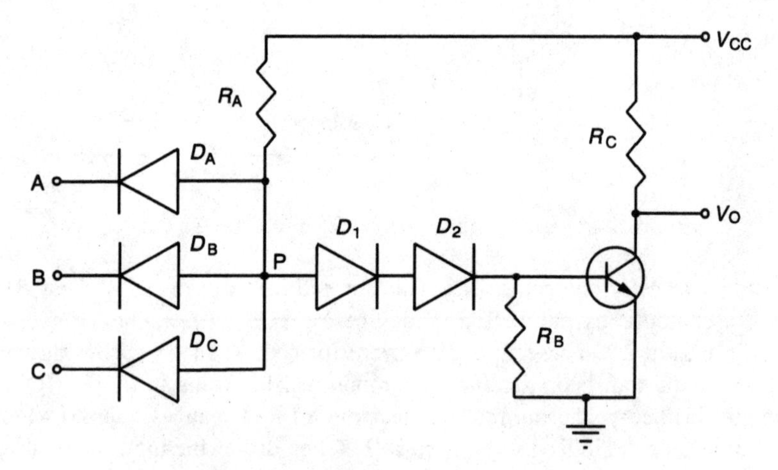

Figure 2.8 DTL three-input NAND gate

0). Two diodes are required to ensure that the transistor is fully turned off when one of the inputs is low.

DTL has good noise margins and fan-out capability, but is rather slow in switching. This is due primarily to two factors. Firstly, when the transistors switch from the on-state to the off-state, the charge stored on the base has to leak through R_B to ground, and the time for this discharge to take place can be comparatively long. The second cause arises when the transistor is driving other gates or wiring. This load can be considered as largely capacitive in nature. When the transistor switches off, this capacitive load has to charge up through R_C, which again can take a long time. Because of these drawbacks, and the fact that integrated circuit technology enables multi-emitter devices to be formed, the DTL configuration has itself been superseded by the TTL configuration. Basically, the input diodes are substituted with a multi-emitter transistor, as shown for example in Figure 2.9, which is the TTL version of the DTL gate shown in Figure 2.8. The operation of the gate is subtly different from the DTL version, however.

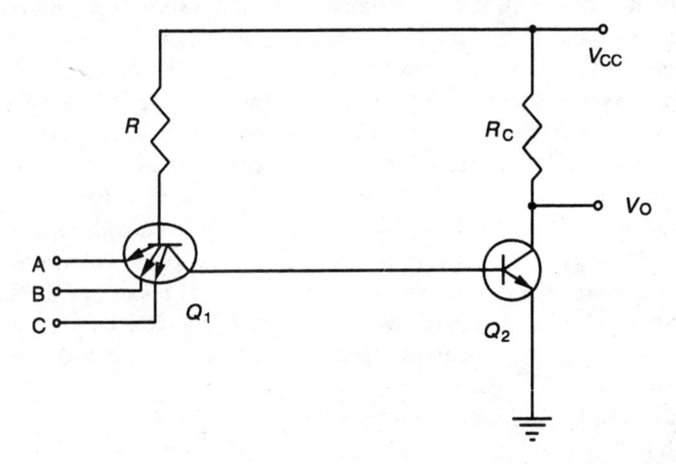

Figure 2.9 TTL three-input NAND gate

Suppose all three inputs are high. Current flows from V_{CC} through the resistor R, forward biasing the base–collector junction of Q_1. The base–emitter junction is however reverse biased. This is the opposite condition to the normal biasing conditions of a BJT and the device is said to be acting in the reverse active mode, that is, in an active mode, but with the emitter and collector terminals reversed. The current gain, β, associated with this reverse mode is very low ($\ll 1$), so the input currents are small and the current flowing into the base of Q_2 is approximately that flowing through R. This current drives Q_2 into saturation and so the output voltage goes low.

Now suppose any of the input voltages is taken low. The corresponding base–emitter junction of Q_1 becomes forward biased and the device switches back into its normal active mode. Q_2 was in saturation, and remains so until the excess charge stored in the base region is removed. In its normal active mode, the β value of Q_1 is fairly high, so the reverse current flowing from the base of Q_2 (β times the current flowing through R) is very high, the base discharges very quickly, and Q_2 switches to the cut-off condition much more rapidly than in the DTL configuration. This has countered one of the timing limitations of DTL. The second can be overcome because the value of R_C can be reduced by about an order of magnitude in this configuration. Further increases in speed and fan-out can be achieved by adding a further push–pull output stage to the gate, at the cost of two more transistors.

The standard TTL technology as described above was introduced in the 1960s and formed the major digital technology for several years. Although it operated at much faster speeds than the earlier RTL and DTL technologies, as time went on it needed to be improved, both in speed and power consumption. The main factor that affects speed of operation in these circuits is the fact that the transistors are driven into saturation when in the on state. When they are then switched off, there is a large amount of charge stored on the base which must be dissipated as quickly as possible. So one way to speed up the circuits is to ensure that the devices are not driven into hard saturation. A way of achieving this is to include a Schottky diode (rectifying metal–semiconductor junction) between the base and collector as shown in Figure 2.10. Schottky diodes can easily be fabricated in standard technologies and do not occupy much substrate area. In normal operation, the forward voltage drop across these devices is around 0.4 V. When the transistor is turned on, the voltage across the emitter–base junction is typically 0.7 V, so the voltage across the emitter–collector (the output voltage) is about 0.3 V, instead of the saturated 0.2 V. So the output voltage is low enough to be considered a logic 0, but the transistor is not in hard saturation and can be switched to cut-off in a much shorter time.

While Schottky TTL has lower gate delays, the power consumption is still fairly high. It is obviously advantageous to have as low a power consumption

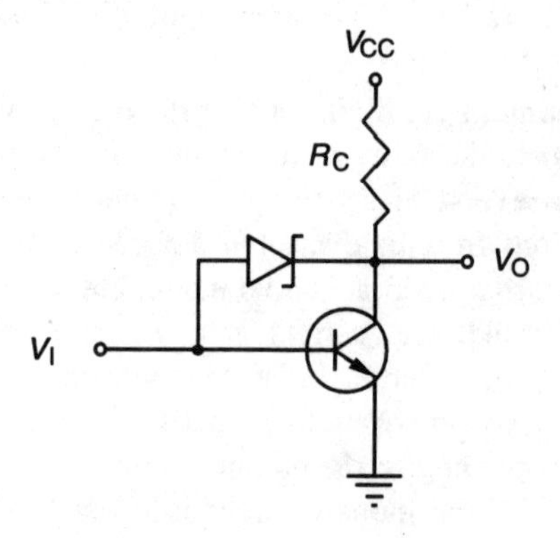

Figure 2.10 Basic Schottky TTL inverter stage

as possible, so a low-power Schottky (LSTTL) technology was developed. There were two main changes in this development. Firstly, higher values of resistors were used, about an order of magnitude greater than in standard Schottky TTL, reducing the power consumption by about a factor of 10. Of course, by increasing the resistances, this will cause an increase in the gate delays (the same argument was used in the change from DTL to TTL), however this increase can be counteracted by the second change, that of substituting Schottky diodes for the multi-emitter input device. The main reason for developing this device was to use the gain of the input transistor to draw a large discharge current from the base of the output transistor and enable fast switching to occur. As this output transistor is not operated in hard saturation, this requirement has been largely eliminated, and diodes can be used at the inputs (effectively going back to a DTL configuration).

With recent improvements in fabrication technologies, reduced component size and lower parasitic capacitances, there have arisen variations of STTL and LSTTL with even lower power consumptions and gate delays, termed advanced Schottky and advanced low-power Schottky (ASTTL and ALSTTL). A comparison of the performance of these various technologies is given in Table 2.1. Delay–power product is a useful figure for assessing overall gate performance, being the product of the propagation delay and the power dissipation. The aim is to achieve as low a value as possible.

Table 2.1 *Comparison of TTL technologies*

	TTL	*STTL*	*LSTTL*	*ASTTL*	*ALSTTL*
Propagation delay (ns)	10	4	10	2.5	4
Power dissipation (mW)	10	16	2	8	1
Delay–power product (pJ)	100	64	20	20	4

2.3.2 ECL

As mentioned in the previous section, the main limitation to the speed of bipolar logic gates is the fact that the devices are operated in or close to saturation, and to switch them to cut-off requires a movement of charge from the base region which takes a certain time to complete. An alternative form of bipolar logic circuit has been developed in which the devices are not operated in saturation, and are inherently much faster in their operation. This circuit is based around a differential pair configuration as shown in Figure 2.11. As the emitters of the transistor pair are joined together, this technology is usually referred to as emitter-coupled logic (ECL).

The differential pair is a configuration that allows current to flow in only one of the transistors at a time, a process termed current steering. One side of

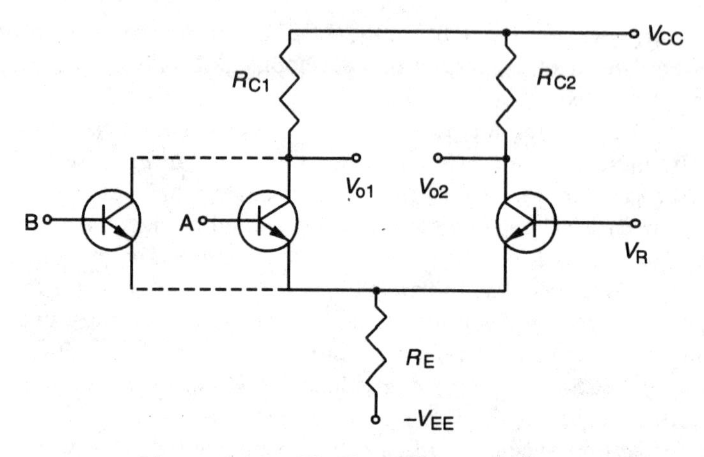

Figure 2.11 Bipolar differential pair

the pair is connected to a reference voltage, while the other side is composed of two or more transistors in parallel. If the base voltage of one of these transistors is raised to a voltage about 0.1 V greater than the reference voltage, then the current is steered through the transistor with this higher base voltage. The current drawn through R_{C1} means that the voltage at V_{o1} is lower than that at V_{o2}. This difference in voltage is of the order of 0.8 V. If all the base voltages on the left-hand side of the pair are less than the reference voltage, then the current is steered through R_{C2} and V_{o2} is lower than V_{o1}. As the swing in output voltage between these states is very small, and none of the devices is operated in saturation, the circuit is very fast.

Figure 2.12 shows the overall circuit for the OR/NOR ECL gate of the 10K family of devices. It consists of three stages: the input differential pair, the voltage reference circuit, and emitter follower output transistors to provide sufficient drive capability. A faster version of the circuit, termed the 100K series, is available. Although the speed of ECL is very fast, the power consumption is relatively high. Typical gate delays are 2 ns for 10K and 0.75 ns for 100K, and the power dissipations are 25 mW and 40 mW, resulting in delay–power products of 50 pJ and 30 pJ respectively. ECL operates from a negative supply, so the supply and logic voltage levels are incompatible with the TTL technologies. A summary of the various supplies and logic levels is given later in the chapter.

2.3.3 MOS

As stated earlier, there are two forms of the basic MOS transistor, n-channel and p-channel, termed nMOS and pMOS respectively. In the former the main current is carried by electrons and in the latter by holes. As electrons have a higher mobility than holes in any given material, the n-channel devices can

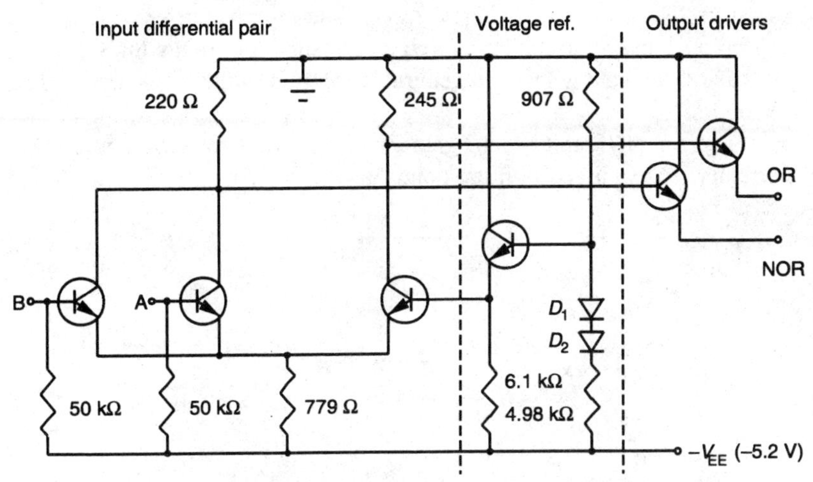

Figure 2.12 ECL OR/NOR gate

operate at higher speeds than the p-channel devices, and are the preferred technology for straightforward MOS circuits.

The input gate structure consists of a small MOS capacitor structure which has a very low value of capacitance, and hence there is little charge storage. Thus the problems of charge storage in the bipolar technology that limited speed and fan-out are not significant in the MOS technology.

The digital operation of these circuits is to operate them at the two extremes of their output characteristics, in the same way as for the bipolar circuits. Remember, however, that in the MOSFET operation there is no equivalent of the hard saturation condition that exists for bipolar circuits, so the voltage corresponding to the logic zero can be relatively high and, perhaps more importantly, the current flow through the device and power dissipated can be quite high. In order to offset this, the drain load resistors would need to be quite large. As large resistors on a monolithic circuit take up a large area, and the cost of a chip is almost directly proportional to area, this would make MOS circuits very expensive to produce.

Fortunately, owing to the structure of the MOS transistor being essentially a voltage-controlled resistor, we can use a transistor configured as an active load. This is generally done by connecting the gate to either the source or drain terminals to achieve the two-terminal resistor. As described earlier, there are two further forms of MOS device: enhancement mode and depletion mode. The first is a 'normally off' device, in that a gate voltage has to be applied in order to create a conducting channel. The second type is 'normally on', in that it has a channel existing even when the gate–source voltage is zero. So in the enhancement mode device, the gate and drain are connected together to form the active resistor; while for the depletion mode device, the gate and source are connected together. Although both types of device can theoretically be used as active loads, the enhancement mode load circuits suffer from noise-margin problems, owing to the low voltage corresponding to the logic 1 state, and also have large gate delays. In general, depletion mode loads are used in nMOS circuits.

Simple inverter, NOR and NAND gate circuits are illustrated in Figure 2.13. These circuits are examined in more detail in Chapter 3.

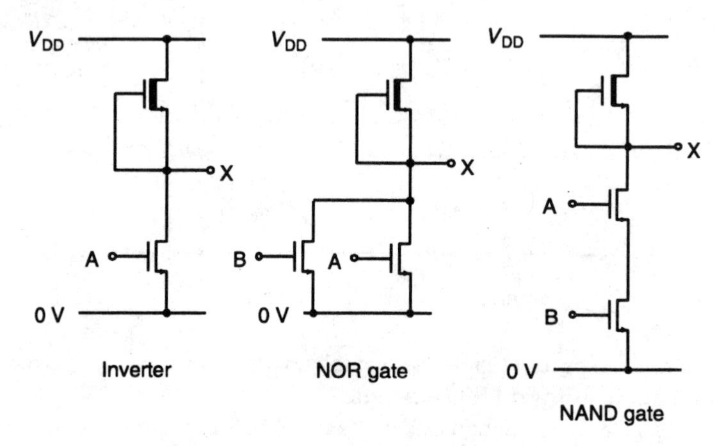

Figure 2.13 Simple nMOS gates

2.3.4 CMOS

The main drawback with the nMOS technology remains its relatively high power consumption, and this is the reason why it is not in greater usage as a monolithic technology. Fortunately, this disadvantage can be almost completely offset by using a combination of nMOS and pMOS devices in complementary pairs. The basic CMOS inverter stage is shown in Figure 2.14, and consists of enhancement mode devices.

The basic operation of the circuit is as follows. With the input at logic 0, approximately zero volts on the gates, the V_{GS} of the n-channel device is

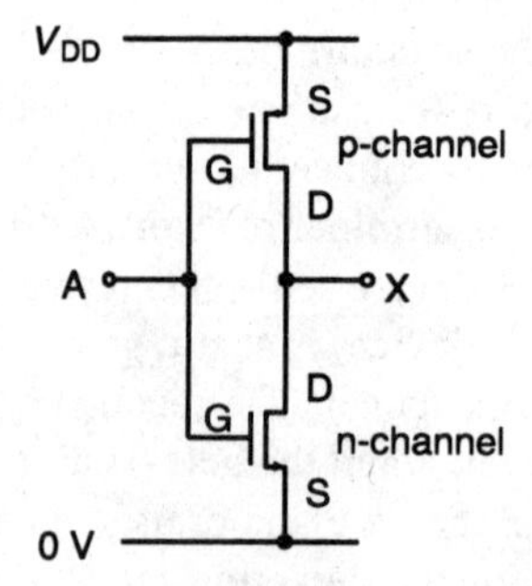

Figure 2.14 Basic CMOS inverter circuit

below its V_T value, and the device is switched off and has a high resistance. The V_{SG} of the p-channel device is approximately equal to V_{DD}, well above the V_T value for this device, so the device is on and has a low resistance. Therefore the potential divider circuit created by the two devices is such that most of the voltage is dropped across the n-channel device, and the output voltage is close to V_{DD}.

With the input voltage at logic 1, close to V_{DD}, the p-channel device is switched off and the n-channel device switched on, so the output voltage is close to zero. Hence the inverting operation is clear. However, the important point is that in the static case, whether the output is logic 0 or logic 1, one of the two devices will be in the switched-off, high-resistance state. Thus the current flow through the potential divider circuit will be virtually zero. The only time when any significant current flow occurs is during the dynamic switching operation. As this transient period is very short, the overall power consumption of CMOS is extremely low and is used wherever power supply is restricted, for instance in portable equipment such as watches, or in air-borne or space applications. If the two transistors are properly matched, the logic transition can occur very sharply at the half supply voltage point, leading to very good noise immunity, and the circuits can be operated at very low supply voltages. These points are also expanded in Chapter 3.

2.3.5 Comparison of digital technologies

Table 2.2 gives a summary of the important parameters of the standard digital technologies described above. This demonstrates the various advantages, disadvantages and limitations of each. It also shows that there is no universal 'best' technology, but the choice depends very much on the particular application. For example, low-power applications demand CMOS, whereas very fast applications demand ECL, etc. The data in the table only gives typical values; there may be some variation between particular manufacturers' products.

Table 2.2 *Comparison of digital technologies*

Logic family	*Supply voltage (V)*	*Power per gate (mW)*	*Propagation delay per gate*	*Maximum clock frequency*	*Maximum logic 0 input (V)*	*Minimum logic 1 input (V)*	*Maximum logic 0 output (V)*	*Minimum logic 1 output (V)*
TTL	5	10	10 ns	35 MHz	0.8	2.0	0.4	2.4
STTL	5	16	4 ns	75 MHz	0.8	2.0	0.5	2.7
LSTTL	5	2	10 ns	40 MHz	0.8	2.0	0.5	2.7
ASTTL	5	8	2 ns	100 MHz	0.8	2.0	0.5	3.0
ALSTTL	5	1	4 ns	50 MHz	0.8	2.0	0.5	2.5
ECL (10K)	−5.2	25	1 ns	500 MHz	−1.48	−1.13	−1.6	−0.98
pMOS	−9	1	4 µs	100 kHz	−4.0	−1.2	−8.5	−1.0
nMOS	5	0.1	100 ns	3 MHz	0.8	2.4	0.4	2.4
CMOS*	3–15	0.5	100 ns	3 MHz	1.5	3.5	0.5	4.5

*CMOS data given for 5 V supply and operation at 1 MHz.

2.4 Analogue circuits

There is a huge variety of circuit functions that are analogue in nature, such as amplifiers, oscillators, filters, comparators, mixers, etc. In terms of monolithic integration, most of these circuits can be, and have been, realized. Although circuit designers try to achieve as much of an electronic system as possible using digital circuitry, the real world is still largely analogue in nature, and most of the incoming and outgoing signals are of this form. Therefore, it is almost impossible to avoid having some analogue functions, if only analogue-to-digital and digital-to-analogue converters interfacing a wholly digital processor.

These circuits may be linear or non-linear in nature, and so make use of the full range of output characteristics of the devices. The primary building block for analogue circuits is the operational amplifier or op-amp. This is a high-gain, high-input impedance, low-output impedance, differential input circuit which is very versatile and can be used in virtually every analogue circuit design. Some details of this design are covered in Chapter 3. Here it is important to highlight the available technologies. Analogue circuits are realizable in both BJT and MOS technologies, and both are used for monolithic integration; CMOS is used almost exclusively for the latter.

Bipolar devices tend to have better power handling and drive capability, operate at higher speeds, have better noise performance and lower associated offset voltages, and are almost essential for reference voltage circuits. CMOS has higher input impedances, and is almost essential for analogue switches, multiplexers and switched capacitor circuits. So the choice of technology has often been based on the overall circuit function and components required.

However, in recent years the problems associated with the processing and design of mixed bipolar–MOS technologies on the same substrate have largely been overcome, leading to a new technology termed BiCMOS. The BiCMOS technology is an extension of the basic CMOS process, in which npn and pnp bipolar structures can be formed. Figure 2.15 illustrates the basic CMOS fabrication process employing a 'p-well' in which the n-channel device is formed. It can be seen that a lateral pnp device could be formed using the p^+ source region of the p-channel device as the emitter, the n^- substrate as the base region and the p^- well as the collector. The base will, of course, be constrained to the substrate potential, but the emitter and collector potentials are arbitrary, so circuit design is not too constrained. However, this lateral pnp device is not very satisfactory because the base is relatively wide, so the device would have a low β value and low speed operation.

A vertical npn device can be achieved using the n^+ source/drain region of the n-channel device as the emitter, the p^- well as the base, and the substrate as the collector. Here the base width can be made much smaller than the lateral device, and also better control of this dimension is possible. The problem with using the substrate as the collector is that in normal operations the

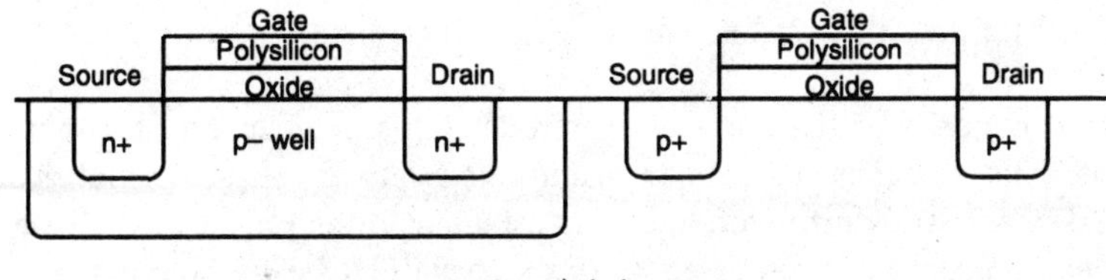

Figure 2.15 Cross-section of p-well CMOS process

base–collector junction should be reverse biased, so the substrate should be connected to the most positive supply voltage (V_{DD}). This can cause many design problems, particularly if pnp devices are also in the circuit, with their own constraints. Additionally, the substrate is only lightly doped, and therefore has a relatively large resistance. As the collector currents are usually quite high, much power can be lost in the substrate; hence the collector is usually connected to an n^+ region formed at the same time as the emitter.

Because of the unsuitability of the lateral pnp device, a vertical pnp device can be formed with extra processing to provide a lower p region which can be brought to the surface of the circuit through a p^+ area, in which the n base width is thin and well controlled.

The processing is complex, and the circuits are therefore relatively expensive, but the advantages with this technology are enormous. As well as providing the best of both worlds, so far as the analogue circuit realization is concerned, it also enables both analogue and digital circuits to be efficiently combined on the same monolithic circuit. This has led to a revolution in integrated circuit design and production, since complete systems and sub-systems are available on a single chip which can interface directly with the outside world and provide very sophisticated processing and control. For example, consider the system illustrated in Figure 2.16. Using the most appropriate technology for each of the sub-systems or components, five separate ICs would have been required. With the use of a BiCMOS process, there is the possibility of integrating the whole system on a single IC.

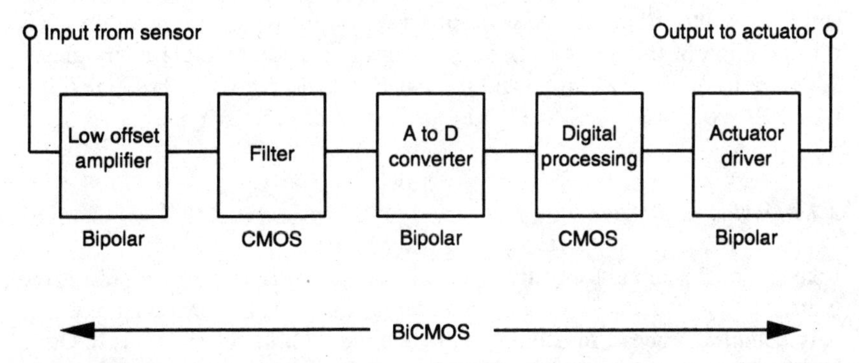

Figure 2.16 Integration of mixed A/D circuits using BiCMOS

2.5 Gallium arsenide and microwave circuits

The vast majority of ICs are manufactured using silicon as the base semiconductor. This is because the silicon material is relatively cheap, the fabrication processes are straightforward and well established, and most of the research and development has been based on this material. However, silicon has drawbacks as the frequency of operation increases. The maximum speed that a device can operate at (switching speed for digital, f_T for analogue) is determined primarily by the time it takes for the charge carriers (electrons and holes) to traverse the device. This in turn is determined by the physical size of the device and the speed at which the carriers travel. The physical size is determined by the mask processing and etching techniques used (see Appendix 1). The present limit for these is of the order of 0.1 μm, and it would require a major advance in processing techniques to reduce this limit significantly.

The speed of carriers is related directly to the applied electric field, the proportionality constant of this relationship being termed the mobility. The higher the mobility, the faster the charge carriers will travel at a particular field strength. Given that there are limits on the dimensions and voltages, increasing field strengths is not an option. The mobility is a relatively constant value for a particular material; for silicon the values are 0.13 $m^2 V^{-1} s^{-1}$ for electron mobility μ_e and 0.05 $m^2 V^{-1} s^{-1}$ for hole mobility μ_h.

To achieve a significant increase in the frequency of operation, we need to move to a different material with a higher mobility. Germanium (Ge) has higher electron and hole mobilities but has disadvantages in processing and circuit performance, for example very high leakage currents, that make it unsuitable for large scale integration. The other alternatives are the so-called compound semiconductors, which comprise a mixture of two or more elements. The usual mix is of elements from group III and group V of the Periodic Table (III–V semiconductors), or alternatively a mixture from groups II and VI of the Periodic Table (II–VI compounds).

There are therefore many combinations of compound semiconductors, the most commonly used being gallium arsenide (GaAs) and indium phosphide (InP). GaAs has now become widespread for both hybrid and monolithic ICs for operation in the microwave range of frequencies. There are many manufacturers in the USA, Europe and the Far East working with GaAs, and also foundries exist throughout the world for custom circuit design.

2.5.1 Device realization

Like monolithic circuits on silicon, the main component in GaAs circuits is the transistor. To fabricate n and p doped regions in the same substrate requires a very complex process. In addition, although the mobility of electrons in GaAs (0.85 $m^2 V^{-1} s^{-1}$) is much higher than that of silicon, the hole mobility in GaAs

($0.04\ m^2\ V^{-1}\ s^{-1}$) is almost the same, so there is no advantage in using holes as the majority charge carrier. Therefore bipolar technologies in GaAs have no great advantages over silicon, except when they are constructed using semiconductor heterostructures, a mix of different compound semiconductors that gives rise to enhanced mobilities. These require even more complex processing of course, and so are very expensive and used only for specialized requirements. The main monolithic GaAs technology is an FET-based one.

A further problem with GaAs is that no native oxide layers can be grown, as there can for silicon. Instead, such layers have to be deposited using, for example, silicon dioxide and silicon nitride to form insulating dielectric layers. An alternative is a material called polyimide which can be spun on and baked to a hard layer in much the same way as photoresist layers.

The consequence of this is that MOSFETs are difficult to fabricate in GaAs, particularly if small physical dimensions are required (which is the case for very high frequency devices). Hence the prime technology is to form the gate of the structure using a Schottky contact. Thus the gate metallization is laid directly on to the channel semiconductor and the capacitance of the reverse-biased diode structure acts as the control for the drain–source current, forming the metal–semiconductor FET (MESFET).

A consequence of the operation of these circuits at microwave frequencies is that the interconnection lines are comparable in length to the wavelength of the signals, and so transmission line effects have to be taken into account in the design. This is manifested in the lines having a characteristic impedance, and wherever there is a change in the impedance there will be a reflection of power. In addition the crosstalk effects, or coupling of signals between nearby lines, become of greater importance. The metal interconnect lines on top of the semiconductor form the classic microstrip transmission line structure. For this to function properly, the dielectric between the line and the groundplane should be relatively loss free, that is of high resistivity. As the GaAs is forming this substrate material, these semiconductor dielectrics should not be highly doped; ideally they should comprise purely intrinsic, undoped material. However, in the growth of the GaAs material there is almost always some residual impurity or natural dopant, so the material is often given a small background dopant level of the opposite type in a process called compensation doping, in order to obtain the highest resistivity possible for the substrate material. Hence these substrates are often referred to as 'semi-insulating' (SI).

Of course, we still need to have a higher doped 'active' layer of GaAs, which must be formed by an epitaxial growth process. It is on this epi-layer that the active devices are formed. However, interconnects and passive components must sit on the SI substrate, so the epi-layer is etched away in these regions, leaving the active components sitting on top of a mesa structure.

In these monolithic circuits there will inevitably be cross-overs of the interconnect lines, so there must be at least two metallizations separated by an insulating layer, with via holes for electrical connection as required. These are usually separate from the gate metallization used for the Schottky contacts, so the most common process will comprise a three-metal layer technology. In addition, there is usually some combination of nitride, oxide and polyimide layers for insulating the metal layers and providing dielectric structures. This relatively complex processing technology obviously has the disadvantage of increased cost, but it does give more scope for the formation of other components that are impractical or unnecessary at lower operating frequencies. For example, inductors can be formed by loop or spiral transmission lines, capacitors by a sandwich of two metal layers with polyimide or nitride acting as the dielectric, and resistors by small mesa sections of the epi-layer. The values of the inductance and capacitance are relatively small (in the nH and pF range respectively), but at microwave frequencies such components have a significant and useful reactance. Hence these components can be used as tuning, matching or d.c./r.f. block elements, as required.

GaAs circuits are certainly a growth area and will continue to be so. The main applications at the moment are in the analogue field, particularly in communications (for example, in satellite receiver front-ends) and radar equipment, but are also seeing increased use in digital circuits as bit and clock rates break through the Gb s^{-1} barrier.

Bibliography

P.E. Allen and D.R. Holberg, *CMOS Analog Circuit Design*, Holt Reinhart & Winston, New York, 1987.

J. Allison, *Electronic Engineering Semiconductors and Devices*, McGraw-Hill, London, 1990.

T.E. Dillinger, *VLSI Engineering*, Prentice-Hall, Englewood Cliffs, New Jersey, 1988.

E.D. Fabricius, *Introduction to VLSI Design*, McGraw-Hill, New York, 1990.

P. Gray and R. Meyer, *Analysis and Design of Analog Integrated Circuits*, Wiley, New York, 1993.

P.J. Hicks, *Semi-Custom IC Design and VLSI*, Peter Peregrinus, London, 1983.

R.S. Pengelly, *Microwave Field Effect Transistors*, 2nd edn, Research Studies Press, Letchworth, Herts, 1986.

C.J. Savant, M.S. Roden and G.L. Carpenter, *Electronic Design*, 2nd edn, Benjamin/Cummings, Redwood City, California, 1991.

A.S. Sedra and K.C. Smith, *Microelectronic Circuits*, 2nd edn, Holt Reinhart and Winston, New York, 1987.

R.S. Soin, F. Maloberti and J. Franca, *Analogue–Digital ASICs*, Peter Peregrinus, London, 1991.

N. Storey, *Electronics – A Systems Approach*, Addison-Wesley, Wokingham, Berks, 1992.

H.J.M. Veendrick, *MOS ICs – From Basics to ASICs*, VCH, Weiheim, Germany, 1992.

Questionss

2.1. From the BJT output characteristics of Figure 2.2, estimate the common emitter current gain $\beta = \left.\frac{dI_C}{dI_B}\right|_{V_{CE\ const}}$ [450]

2.2. From the MOSFET output characteristics of Figure 2.5, estimate the mutual transconductance

$$g_m = \left.\frac{dI_D}{dV_G}\right|_{V_{DS\ const}}$$ [1.5 mS]

2.3. Sketch the circuit of an RTL NAND gate.

2.4. From the data in Table 2.2, calculate the power–delay product for each logic family. Which technology has the best and worst values?
[100 pJ, 64 pJ, 20 pJ, 20 pJ, 4 pJ, 50 pJ, 4000 pJ, 10 pJ, 50 pJ]

3 Transistor-Level Design
What are the building blocks?

3.1 Introduction

In the previous chapter we explored the various circuit technologies available to the IC designer and also touched briefly on how the devices could be formed into the next layer of the hierarchy, for example digital gates. In this chapter we shall look in more detail at the design of these building blocks, both for digital and analogue circuits. This will cover not only the basic topologies and design approaches, but also a consideration of power, loading and layout of the circuit elements in real ICs.

3.2 Digital circuits

Any digital circuit, however complex, can be broken down into a number of small building blocks called gates which perform simple Boolean operations. It is this level of circuit hierarchy that we shall examine here. As was described in Chapter 2, there are four main technologies available for digital circuits: TTL and ECL based on BJTs, and nMOS and CMOS based on FETs. We shall look at the detailed operation of certain gates in each of these technologies in order to highlight the important design considerations, which also influence such aspects as power dissipation, loading effects, circuit layout and testing.

One principal characteristic of the gates we shall be looking at in this chapter is that of the voltage transfer curve, that is a plot of V_{out} against V_{in}. It will be seen that in general there are no specific values of voltage for the two logic levels of 1 and 0. The actual voltages at input or output that correspond to the logic values fall within a range of acceptable voltages, the values of which depend on the circuit technology being used. The range of input voltage values for the two levels must be greater for the input logic levels than the output levels. This is because the output signal, when used to drive a further gate, will inevitably degrade as a result of noise, losses and cross-talk. This leads to the definition of noise margins, as illustrated in Figure 3.1. Here the acceptable range of voltages is illustrated for the input and output logic levels. The

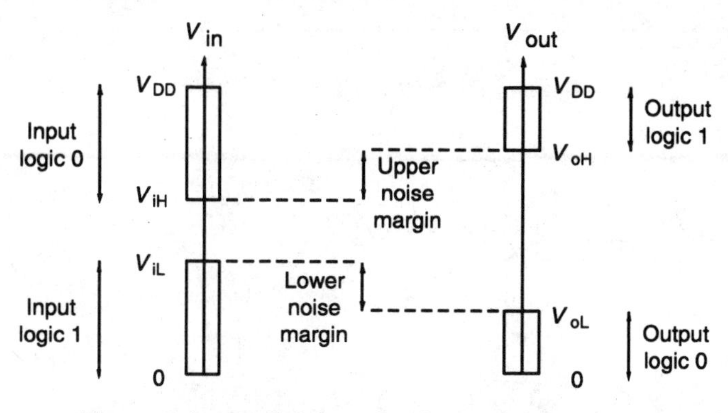

Figure 3.1 Definition of noise-margin voltages

difference between the lower limits for the logic 1, and the upper limits of the logic zero define the two noise margins.

The other principal characteristic that will be considered is the loading effect of one gate driving a number of subsequent gates. The loading may be a d.c. effect, in terms of the maximum current that a gate can supply before the output voltage degrades beyond the limits indicated in Figure 3.1, or it may be limited by the capacitive loading of subsequent stages which limits the speed at which the voltage can change. In any event, the maximum number of similar gates that a gate can drive before the output voltage degrades beyond the defined limit is termed the fan-out of the gate.

3.3 TTL gates

TTL gates are in practice rarely used in circuits more complex than small scale integration (SSI). However, the detailed operation of a standard TTL gate is explained here, partly for completeness and partly because bipolar technologies are still evolving. In particular the development of heterojunction bipolar transistors (HBTs) with very high operating speeds has ensured the future of digital bipolar technologies.

The operation of a very basic TTL gate was described qualitatively in the previous chapter. In fact the circuit described there is very sensitive to loading effects, and is not a practical realization. The standard TTL inverter stage is shown in Figure 3.2: it can be seen that this circuit is similar to that of Figure 2.9, but that the former output transistor is now a phase-splitting transistor driving a two-transistor 'totem-pole' output stage. We shall examine the operation of this gate quantitatively and derive its transfer characteristics. For this analysis we shall assume that the transistors are identical, have a V_{BE} value for initial turn on of 0.6 V, and values of V_{BE} and V_{CE} in saturation of 0.7 V and 0.2 V respectively. These are generalizations and the actual transis-

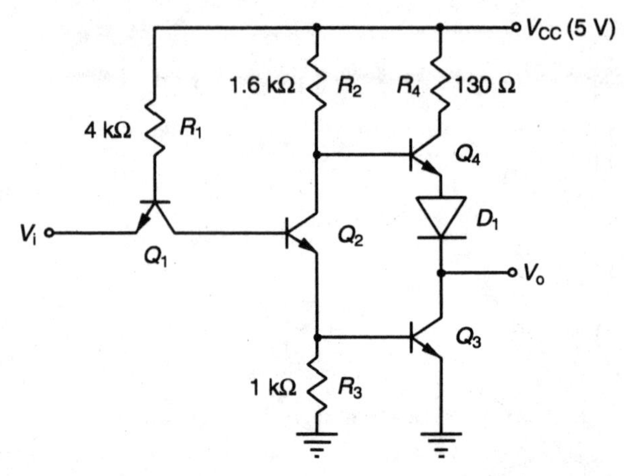

Figure 3.2 Basic TTL inverter stage

tor behaviour will be more complex than this, but these approximations are good enough for the present purposes.

With $V_i = 0$, the base current flowing through R_1 ensures that Q_1 is turned hard on into the saturation region and there is no current flow into the base of Q_2 which is therefore turned off. No current flows through the emitter of Q_2 so the voltage across R_3 is virtually 0 and Q_3 is also turned off. Sufficient current flows into the base of Q_4 for this device to be turned on. The voltage at the output is the value of V_{CC} less the drops across D_1, V_{BE} (Q_4) and R_2. For a 5 V supply this value is typically 3.6 V.

Increasing the value of V_i, Q_2 will start to turn on when its V_{BE} value, given by $V_i + V_{CEsat}$ of Q_1, exceeds about 0.6 V, that is when V_i is approximately 0.4 V (point A in Figure 3.3). As V_i is further increased, more of the base current into Q_1 is passed into the base of Q_2 which is now operating within its linear region. More current is drawn through Q_2 and hence through R_3, so the V_{BE} of Q_3 starts to rise. Q_4 and D_1 remain conducting, but as more current is drawn

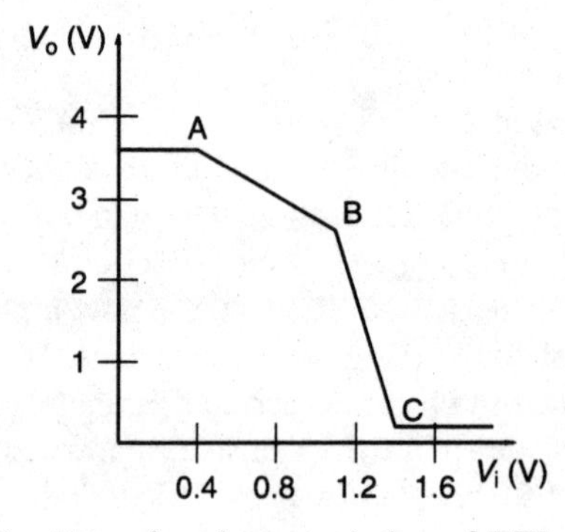

Figure 3.3 Transfer characteristics of TTL inverter stage

through R_2, the voltage drop across this element increases and so V_o decreases in a linear fashion. This continues until Q_3 starts to turn on, which will be when its V_{BE} exceeds 0.6 V, that is when $V_i = V_{BE3} + V_{BE2} - V_{CE1} = 0.6 + 0.7 - 0.2 = 1.1$ V. The current through R_3 is about 0.6 mA, which is approximately the same current as through R_2. Therefore the voltage drop across R_2 is approximately 1 V and V_o is $5 - 1 - 0.7 - 0.7 = 2.6$ V (point B).

As V_i is further increased, Q_3 operates in the active (linear) mode, as do Q_2 and Q_4, the whole circuit acting as a linear amplifier. This will continue until Q_3 saturates, and V_o will then become fixed at V_{CEsat} for Q_3, or about 0.2 V. This will happen when V_{BE} for Q_3 is approximately 0.7 V. Beyond this point, yet more of Q_1's base current will be drawn into the base of Q_2, and Q_1 passes from saturation into the inverse active mode. At the point Q_3 becomes saturated, the V_{CE} value of Q_1 is approximately 0, so V_i is equal to $V_{BE2} + V_{BE3} = 1.4$ V (point C). These overall piecewise linear transfer characteristics are plotted in Figure 3.3.

It can be seen from this that the output voltage of a TTL gate corresponding to the logic 1 is far from the rail or supply voltage of 5 V. In addition, this analysis has been done without consideration of the loading effects of any subsequent gates connected to this output. In practice the current driving capability of the gate is quite high, and usually around 20 gates can be driven before the minimum logic 1 output voltage of 2.4 V is reached (Table 2.2), which defines the fan-out. If more gates need to be driven than the fan-out allows, a buffer gate must be used whose output follows the input, but which has a higher output drive capability and therefore fan-out than a normal gate.

The basic inverter stage of Figure 3.2 can be extended to other gate functions. For example, a NAND function can be achieved by using a multi-emitter device for transistor Q_1, and a NOR function by paralleling the phase splitter transistor Q_2. A 'universal' AND-OR-INVERT TTL gate is shown in Figure 3.4 which provides a general logic function that can be used to realize any of the main logic functions by the proper selection of the input functions.

The layout of BJTs and the mask layers required for fabrication are relatively straightforward for TTL, and are outlined in Appendix 1. The size of the transistor structures, in particular the width of the base, tends to determine the speed of the devices. This is not really a critical design consideration as TTL is mostly used as a general-purpose, relatively slow speed technology. In addition, a large number of on-chip resistors are required in the circuits. These are formed by patterning areas of doped semiconductor, but for larger values of resistance these have to be relatively long and thin. Although they are often formed in a meander pattern to optimize the circuit area used, even so the chip area dedicated to the resistors often exceeds that dedicated to the transistors, so little is gained by reducing the size of the transistor structures. As a result, the TTL technology is rarely used at LSI or VLSI levels of integration, but has found its niche in low-cost SSI implementations for small amounts of random

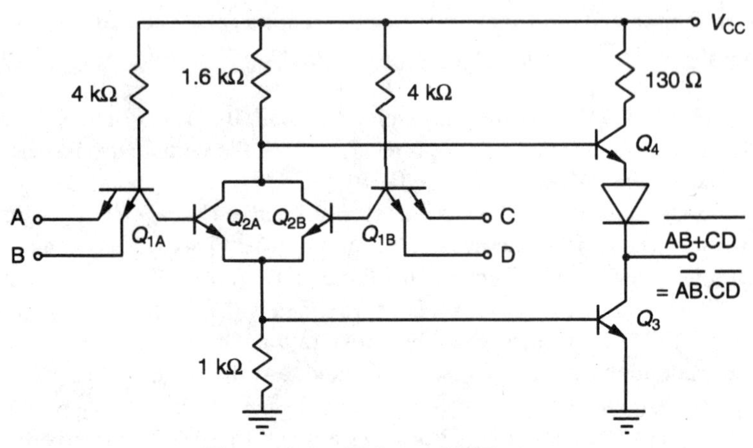

Figure 3.4 Universal TTL gate

or glue logic at the board level of circuit or system realization. Very little custom design is done at the chip level.

3.4 ECL gates

The basic working of the ECL technology was described in Chapter 2 and an illustration of the circuit of a typical gate was given in Figure 2.12.

As a first stage to the quantitative analysis of this circuit, we shall calculate the reference voltage V_R using the circuit shown in Figure 3.5. By assuming that the voltage drops across the two diodes and the base–emitter junction of the transistor are 0.7 V, and that the current gain (β) of the transistor is reasonably high (say 100, although the actual value is not critical for the calcula-

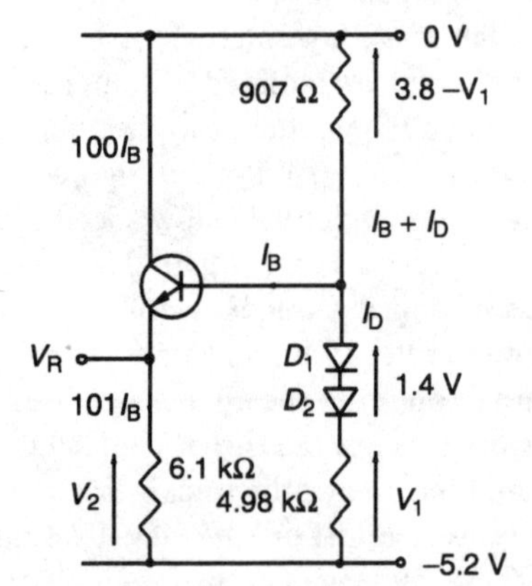

Figure 3.5 ECL reference circuit

tion), the voltages and currents are as shown. We have four unknowns, and therefore require four equations to solve for the value of V_R, as given below:

$$(3.8 - V_1) = 907(I_B + I_D)$$

$$V_1 = 4.98 \times 10^3 I_D$$

$$V_2 = 6.1 \times 10^3(101 I_B)$$

$$V_2 = V_1 + 0.7$$

By eliminating, in turn, I_D, I_B and V_1, it can be calculated that $V_2 = 3.91$ V, so the reference voltage $V_R = -1.29$ V. It can be seen that this value is roughly half way between the minimum logic 1 level (−1.48 V) and the maximum logic 0 level (−1.13 V) for this technology (Table 2.2). We shall return to the significance of this later in the analysis.

The next step is to look at the differential amplifier stage, which is the heart of the gate. Refer to Figure 3.6, which shows the rest of the ECL gate, including the two emitter follower output drivers. These are to provide a low-output

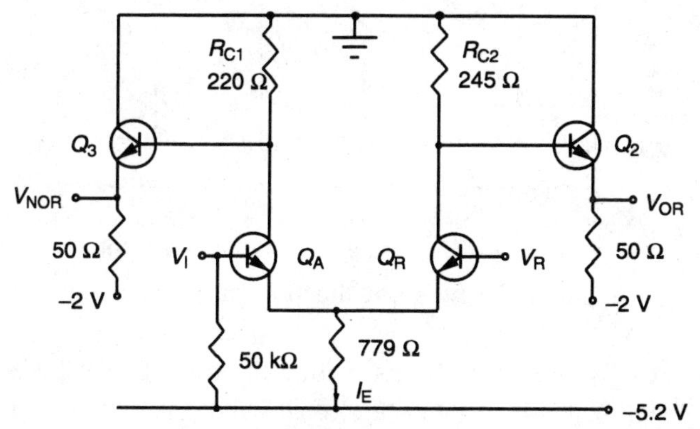

Figure 3.6 ECL differential pair and output drivers

resistance, high-current driving capability for the output signals. They do not have on-chip loads – the 50 Ω resistance to the −2 V supply indicates a properly terminated load at the end of a high-frequency transmission line, which is generally used for interconnections in ECL circuits. The basic operation of the differential pair is to act as a current splitter, the ratio of the currents through R_{C1} and R_{C2} being determined by the relative values of the base voltages. The base voltage of one side is fixed at the reference voltage. If the base of any of the transistors on the other side of the circuit exceeds this voltage by a small amount, virtually all of the current is steered through that side of the circuit, and the voltage drop across R_{C1} is much greater than across R_{C2}. The ratio of currents is related to the ratio of the base voltages by

$$V_{BE}(Q_R) - V_{BE}(Q_A) = \frac{kT}{e} \ln\left(\frac{I_E(Q_R)}{I_E(Q_A)}\right)$$

where k is Boltzmann's constant, e is the charge on an electron and T is the absolute temperature.

For the purposes of analysing the operation of the gate and deriving the transfer characteristics, we shall define the limits of the transfer as being when 99 per cent of the current is directed through one or other of the transistors, so the ratio of the currents is 99, and the difference in base voltages at room temperature is 115 mV.

First of all we shall look at the transfer curve for the OR output, that is V_{OR} against V_I. The general shape is as shown in Figure 3.7, and we are interested

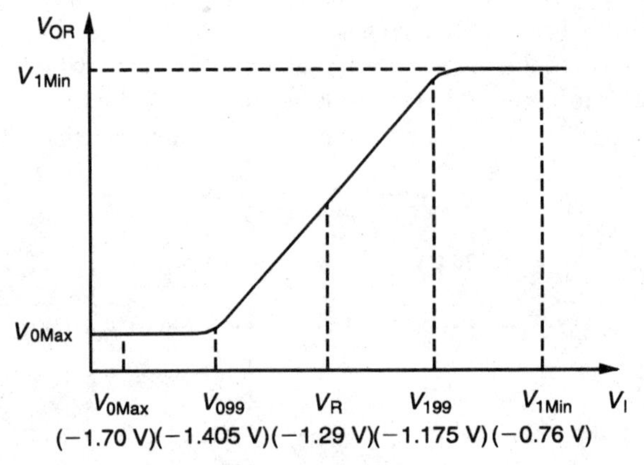

Figure 3.7 ECL OR function transfer curve

in determining the values of V_I that define the maximum logic zero value (V_{0Max}) and minimum logic one value (V_{1Min}). As this is a non-inverting transfer curve, these are defined at the points when $V_{OR} = V_I$. To calculate these we will assume that one or other of the transistors of the differential pair is completely cut off, so all of I_E flows through one device or the other. Also in Figure 3.7 are shown the 99 per cent current points (V_{099} and V_{199}), and the significance of these points will be discussed later.

It is assumed that transistors Q_A and Q_R are an identical pair at the same temperature. (With an IC realization and close proximity of the two devices on the chip, this is a reasonable assumption.) Under these conditions, with V_I equal to V_R, equal currents flow through the two transistors and the transfer characteristic is antisymmetrical around this point. As was calculated earlier, the 99 per cent points are within ±115 mV of this point, that is $V_{099} = -1.29 - 0.115 = -1.405$ V and $V_{199} = -1.29 + 0.115 = -1.175$ V.

To determine the value of V_{0Max} we have the condition that Q_A is hard off and Q_R carries all the current I_E, which is given by

$$\frac{V_R - V_{BE}(Q_R) + V_{EE}}{R_E} = \frac{-1.29 - 0.7 + 5.2}{779} = 4.1 \text{ mA}$$

Therefore $V_{0Max} = V_C(Q_R) - V_{BE}(Q_2) = -4.1 \times 0.245 - 0.7 = -1.70$ V.

To calculate the value of V_{1Min} we know that Q_R is turned off, and all of I_E flows through Q_A. In this case all the current through the 245 Ω resistor flows into the base of Q_2 which is turned on. The part of the circuit of interest is shown in Figure 3.8. Assuming a β value of 100, this results in a base current of 245 μA, and V_o is given as (−) the voltage across the 245 Ω resistor (0.060 V) − $V_{BE}(Q_2) = -0.76$ V.

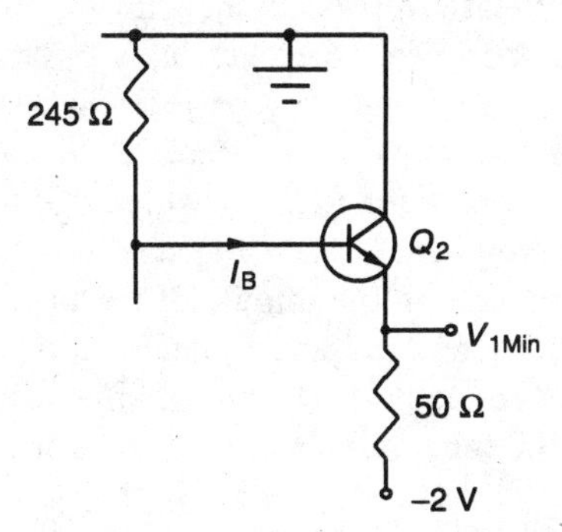

Figure 3.8 ECL output driver

These various calculated values are shown in Figure 3.7. It can be seen from this that the transfer curve is approximately symmetrical in terms of V_I, with V_{1Min} about 0.5 V above V_R and V_{0Max} about 0.4 V below. The values between V_{0Max} and V_{099} and also between V_{1Min} and V_{199} define the noise margins. These two noise margins are approximately equal, and in addition the mid-point of the output voltage swing (corresponding to $V_I = V_R$) is also approximately equal to V_R. These conditions are optimal from the point of view of the noise margins, and this is the reason why the supply voltage is chosen as −5.2 V and the reference voltage is about −1.3 V.

The transfer curve of the NOR output is similar in shape to the OR curve, except of course that it is an inverting function. The transition values are very similar, as the circuits analysed previously are almost identical, except for the slight difference in the value of the collector resistors.

Fan-out restrictions in TTL gates are limited by the d.c. current driving capability of the output stage. In ECL the gates are usually operated at much higher speeds, in other words the component frequencies that comprise the digital square waves are very much higher, hence the need for transmission line designs for the interconnects to maintain the same velocity of these components and make sure that the square wave is not 'spread' in time and distance. The point is that, at these frequencies, the d.c. current capability becomes a secondary factor – as far as fan-out is concerned, the capacitive loading of subsequent gates is more important. The *RC* time constant of the driven circuit limits how fast the square wave can rise or fall to the correct voltage corresponding to a 1 or a 0. The more gates being driven, the larger the value of capacitance, and the larger the time constant. Eventually the rise time will become so large that it exceeds the pulse width and the circuit will no longer function properly. In practical ECL circuits the fan-out from current driving may be as high as 90, whereas the capacitive loading limit may only be 10.

As it is based on the differential 'current-steering' amplifier, the power consumption of an ECL gate is approximately the same in either state, unlike TTL. There is some variation in the current drawn by the output stages as the device switches, and sometimes the power lines to the output stages are decoupled from the differential and voltage reference circuits, so that spikes introduced on the output supply do not affect the operation of the basic gate. This is particularly important for ECL in which the transition voltages and noise margins are relatively small. We have seen that the reference voltage is arranged to be approximately in the middle of the high- and low-voltage values, and so the noise margins are approximately equal. In addition, the reference circuit is designed such that with temperature variations of the components, the noise margins remain reasonably constant.

3.5 nMOS gates

We shall first look in a little more detail at the operation of MOS transistors in order to provide a basis for the circuit design techniques in this and subsequent sections. Although nMOS circuits only employ n-channel devices, we shall also cover the p-channel device, as this is relevant to the sections on CMOS circuits.

The important characteristics of a MOS device are the output characteristics (I_D against V_D) and the threshold voltage of the device, V_T. The output characteristics of the n-channel enhancement and depletion mode devices and the p-channel enhancement mode device are shown in Figure 3.9. Typical values of drain current are shown on these plots, the value being in fact a function of the geometry of the gate. If the gate length (distance between source and drain areas) is L and the gate width is W (that is, the gate area of MOS structure = WL), then the current is directly proportional to W/L (see the following subsection). The gate length also plays an important role in the speed of the transistor, as this depends on the transit time of the carriers from source to drain. The area WL determines the input capacitance of the device, which relates to loading effects and fan-out limitations. Thus the design of a particular gate size has many factors, and we shall examine these in more detail. The values

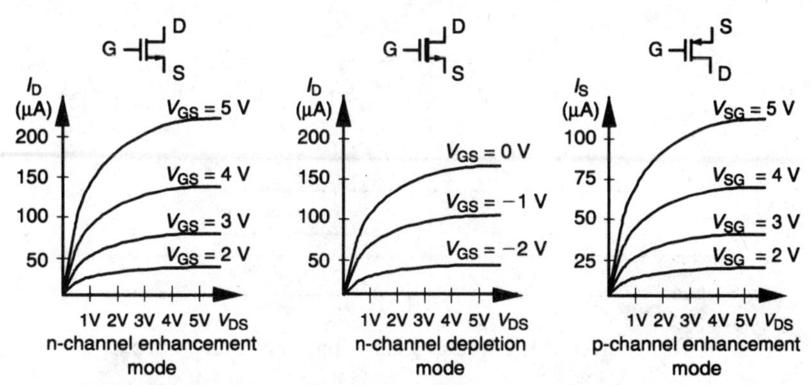

Figure 3.9 Output characteristics of various MOSFETs

shown in Figure 3.9 are for gate sizes of $W = L$ in each case, and threshold voltages of 1 V for the enhancement devices and -3 V for the depletion device.

3.5.1 Derivation of voltage–current relationships

We shall be examining in detail the operation of gates formed by these transistors, so it is useful to know the mathematical relationships between the voltages and currents illustrated in Figure 3.9. We shall derive these relationships from first principles based on the operation of the FET as described qualitatively in Chapter 2.

Linear region

This is where the gate voltage is above the threshold voltage in the enhancement mode device, but the drain–source voltage is sufficiently low for pinch off not yet to have occurred, that is $V_{DS} < V_{GS} - V_T$. The voltage across the capacitive structure of the gate varies along the length of the gate, from V_{GS} at the source end to $V_{GS} - V_{DS}$ at the drain end. The variation is linear and can be considered as an infinite series of capacitor structures with different voltages on the plates. The elemental capacitor structure is shown in Figure 3.10. Here the thickness of the oxide is D, the width of the capacitor W, the elemental length dx and the permittivity of the oxide ε. From the simple parallel plate analysis of this structure, the capacitance of the structure is given by

$$C = \frac{W\,\varepsilon \mathrm{d}x}{D}$$

The voltage in excess of the threshold voltage ('useful' gate voltage) is given by

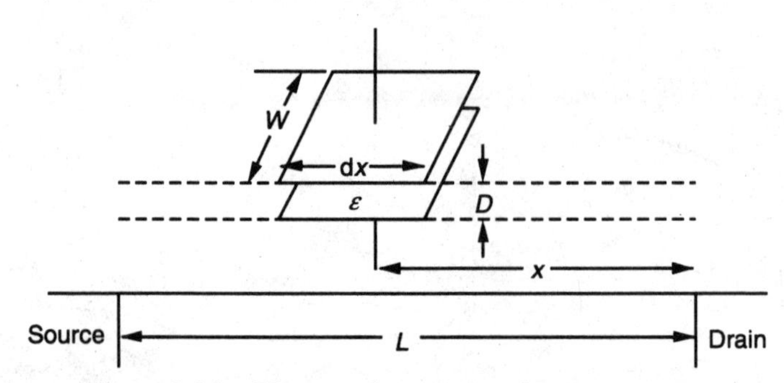

Figure 3.10 Elemental gate capacitance structure

$$v = V_{GS} - V_{DS} + \frac{x}{L} V_{DS} - V_T$$

So the charge on this elemental capacitor is

$$q = Cv = W \frac{\varepsilon \mathrm{d}x}{D} \left(V_{GS} - V_{DS} + \frac{x}{L} V_{DS} - V_T \right)$$

To obtain the total charge on the structure, we integrate across the length of the gate, that is with respect to x:

$$Q = \int_0^L \frac{W \varepsilon}{D} \left(V_{GS} - V_{DS} + \frac{x}{L} V_{DS} - V_T \right) \mathrm{d}x$$

$$= \frac{W \varepsilon L}{D} \left[(V_{GS} - V_T) - \frac{V_{DS}}{2} \right]$$

This is the same value as the charge induced in the channel, which is derived by flow of current from drain to source. This charge is given by tI_D where t is the time taken for the charges (electrons) to flow across the channel. The velocity of electron flow is directly proportional to the electric field, E, inducing the electron movement, and the constant of proportionality is called the mobility, μ_e. Hence t is given by

$$t = \frac{L}{\mu_e E} = \frac{L^2}{\mu_e V_{DS}}$$

Now the drain current is given by

$$I_D = \frac{Q}{t} = \frac{W \varepsilon \mu_e}{LD} \left[(V_{GS} - V_T) V_{DS} - \frac{{V_{DS}}^2}{2} \right]$$

Saturation region

This is the part of the curve beyond pinch off when $V_{DS} \geq V_{GS} - V_T$. The current through the device no longer increases as V_{DS} is increased, and thus saturates at the value corresponding to the pinch-off voltage. Substituting $V_{DS} = V_{GS} - V_T$ into the above equation, we obtain

$$I_D = \frac{W\varepsilon\mu_e}{2LD}(V_{GS} - V_T)^2$$

Notice here that we have the result that was quoted earlier: the drain current is proportional to the ratio *W/L*, and also in saturation the current is proportional to the square of the gate voltage.

p-channel MOSFETs

We can perform a very similar analysis for the p-channel device, and for the resistive region the equation becomes

$$I_S = \frac{Q}{t} = \frac{W\varepsilon\mu_h}{LD}\left[(V_{SG} - V_T)\,V_{SD} - \frac{{V_{SD}}^2}{2}\right]$$

where μ_h is the mobility of the holes, as these are the majority carriers in the p-type channel. The polarity of the voltages and currents are reversed from the n-channel case.

For the saturation region the corresponding equation is

$$I_S = \frac{W\varepsilon\mu_h}{2LD}(V_{SG} - V_T)^2$$

3.5.2 The nMOS inverter

The nMOS inverter gate is based on the operation of the transistor as a switch between the 'on' and 'off' states with some load element as shown in Figure 3.11. The load may be a simple resistor, but to obtain a sufficiently low output voltage in the zero logic state the resistance would be too high to make it

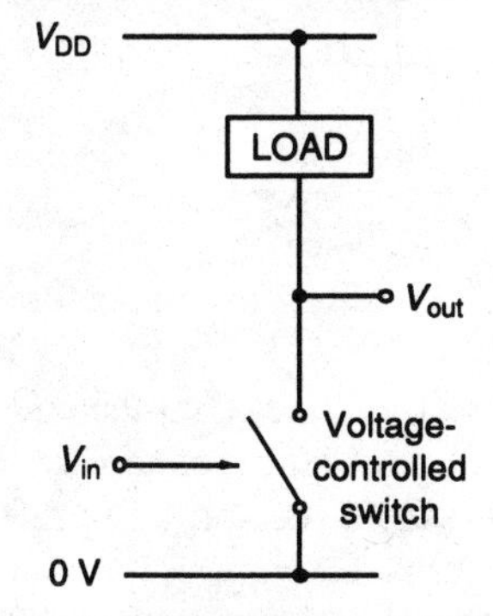

Figure 3.11 Basic form of nMOS inverter

practical for an IC realization. On-chip resistors generally occupy much more area than a transistor (often by a factor of hundreds), and as the price of a chip is directly proportional to its area, cost dictates as small a circuit area as possible.

Thus the alternative is to have another transistor acting as an active load. There is still the option within the nMOS process of using either enhancement mode or depletion mode devices as the active load, and we shall examine these two cases in detail to consider the advantages or disadvantages of each approach.

Enhancement-mode load

As was mentioned in Chapter 2, having the 'normally-off' enhancement mode device as the inverter load, the gate and drain of the load are both connected to the positive supply line, V_{DD}. The output load for the inverter will generally be other transistor gates, which as we have seen are largely capacitive in nature. Hence the operation can be analysed as though the inverter is driving a purely capacitive load. The circuit is therefore as shown in Figure 3.12. Note the subscripts ℓ for the 'load' transistor T_2, and d for the 'driver' transistor T_1. We shall assume in this analysis that the threshold voltage of the two devices is the same, V_T.

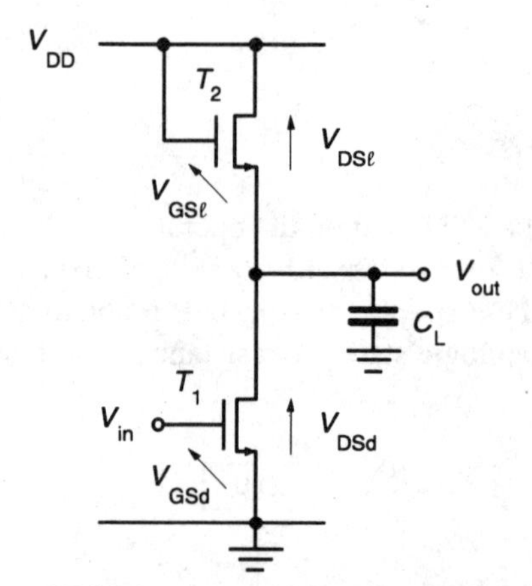

Figure 3.12 nMOS inverter with enhancement mode load

As $V_{GS\ell}$ and $V_{DS\ell}$ are always the same, then the condition $V_{DS} > V_{GS} - V_T$ is always fulfilled and the load transistor must always be operating in the saturation region or be below threshold (cut-off). In practice, if the load were to be in cut-off, it would no longer be acting as a resistive load, and it could provide no current flow to a load. So what happens is that the device is always conducting or on the point of conducting, and the minimum value of $V_{GS\ell}$ and $V_{DS\ell}$ is the threshold voltage, V_T. This limits the maximum output voltage that can be delivered by the gate. The value of the supply voltage V_{DD} is typically 5 V and the threshold voltage typically 1 V. However, this value is affected by the so-called 'body effect', as the substrate of the chip is usually connected to ground, so for the load transistor there is a fairly high voltage between the channel and the substrate which has the effect of increasing the threshold voltage by typically 80 per cent. The actual amount is determined by the particular geometry and doping conditions, but varies as the square root of the source–substrate voltage (hence for the grounded source driver transistor, no correction is necessary). For this circuit, V_{out} for the logic 1 state is approximately 3.2 V.

This will occur when the input voltage V_{in} is less than the threshold voltage for T_1 (1 V). Once V_{in} starts to rise above this value, T_1 conducts a current, the same drain current flowing through the two transistors. V_{out} will drop with increasing V_{in}, and both transistors are now in the saturated region. Increasing the input voltage causes the output voltage, V_{DSd}, to drop below the point $V_{DSd} = V_{GSd} - V_T$, and the driver transistor moves into its resistive region. The two channel resistances form a potential divider circuit from which V_{out} is derived. We must therefore design the *W/L* ratios of the two transistors so that the required output voltage corresponding to logic zero is achieved, and also so that the transfer curve is such that the circuit has reasonable noise margins.

We know that the typical logic 1 voltage is 3.2 V, so we shall use this as our input voltage, assuming the circuit is driven from a similar configuration. The output voltage we want, corresponding to logic 0, must be at least below V_T. In fact, to achieve a reasonable margin, the value is usually designed to be about $0.3V_T$, or 0.3 V. So we have $V_{GSd} = 3.2$ V, $V_{DSd} = 0.3$ V, which we can put into the resistive region equation for drain current:

$$I_{Dd} = \frac{\varepsilon\mu_e}{D}\frac{W_d}{L_d}\left[(3.2 - 1.0)\,0.3 - \frac{0.3^2}{2}\right] = 0.615\,\frac{\varepsilon\mu_e}{D}\frac{W_d}{L_d}$$

This will be the same current that flows through T_2, which has $V_{GS\ell} = V_{DS\ell} = 4.7$ V. In this case the source–substrate voltage is very low and the body effect can be ignored, so $V_{T\ell} = 1$ V. Applying the equation for the drain current in the saturated region:

$$I_{D\ell} = \frac{\varepsilon\mu_e}{2D}\frac{W_\ell}{L_\ell}(4.7 - 1.0)^2 = 6.845\,\frac{\varepsilon\mu_e}{D}\frac{W_\ell}{L_\ell}$$

Therefore the ratio of the *W/L* geometries is given by

$$\frac{(W_d/L_d)}{(W_\ell/L_\ell)} = \frac{6.845}{0.615} = 11.1$$

These values are still considered as ratios, the absolute values of the transistor structures depend on a number of other factors. In general, the size of the circuits will be kept to a minimum for speed and cost reasons, although this is limited by the particular fabrication processes involved (see Appendix 1). Often designs are kept in terms of a minimum resolution limit, λ, the actual value of which can be altered from process to process without having to redesign the circuit. In general, long, thin shapes are avoided, so the W/L ratio of 11 would be shared between the two transistors, thus for example $W_d/L_d = L_\ell/W_\ell = 3.3$. There may be other design constraints such as speed or capacitive loading that will affect the relative ratios.

Depletion-mode load

Using the depletion mode transistor, which is the 'normally-on' device, as the active load, the gate and source are connected together as shown in Figure 3.13. In this case $V_{GS\ell} = 0$ under all conditions. The threshold voltage for a depletion mode device is the gate voltage that suppresses the implanted channel, and is negative, so the depletion mode load transistor is always on.

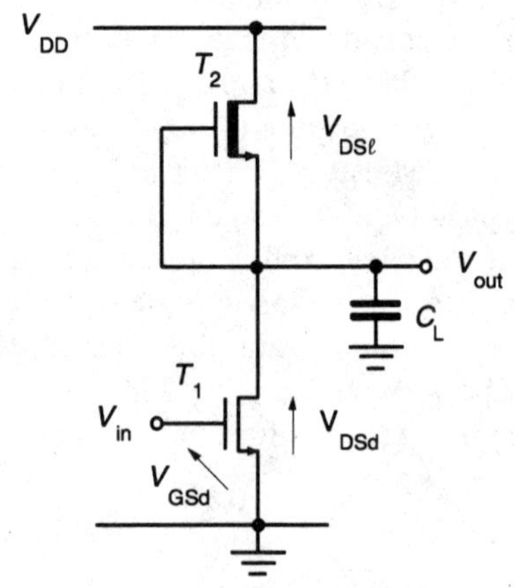

Figure 3.13 nMOS inverter with depletion mode load

When V_{in} is low, less than the threshold voltage of T_1, this transistor is cut off, and only a very small leakage current flows through this device. This can easily be supplied by the load transistor, which has the same I_D value. The voltage drop $V_{DS\ell}$ is also very small, so the output voltage under these conditions is very close to the supply voltage, V_{DD}. This is the primary difference between the depletion-mode load circuit and the enhancement-mode circuit which cannot supply such a high voltage for the logic 1 condition.

Therefore when we consider the behaviour of the inverter in the other switched state, we can use a V_{in} value of 5 V. This is in excess of the threshold voltage of the driver transistor which is therefore turned on. Again the two transistors form a potential divider circuit, so we have to design the transistor sizes to obtain a suitable output voltage for the logic 0 value. We also take this to be 0.3 V. Thus we have $V_{GSd} = 5$ V and $V_{DSd} = 0.3$ V, which is less than $V_{GSd} - V_T$, and the device is operating in the resistive region. So again we consider the equation for I_D in this region:

$$I_{Dd} = \frac{\varepsilon\mu_e}{D}\frac{W_d}{L_d}\left[(5.0 - 1.0)\,0.3 - \frac{0.3^2}{2}\right] = 1.155\,\frac{\varepsilon\mu_e}{D}\frac{W_d}{L_d}$$

This is once more the same drain current for the load transistor, which again has a $V_{DS\ell}$ of 4.7 V, but in this case $V_{GS\ell} = 0$, $V_{DS} > V_{GS} - V_T$ and so this device is in saturation. Applying the relevant formula for drain current:

$$I_{D\ell} = \frac{\varepsilon\mu_e}{2D}\frac{W_\ell}{L_\ell}\left[0.0 - (-\,3.0)\right]^2 = 4.5\,\frac{\varepsilon\mu_e}{D}\frac{W_\ell}{L_\ell}$$

In this case the ratio of the *W/L* geometries is given by

$$\frac{(W_d/L_d)}{(W_\ell/L_\ell)} = \frac{4.5}{1.155} = 3.9$$

It can be seen from the above analysis that the main difference between the two active load realizations is that for the enhancement mode load, the output voltage can never be more than $V_{DD} - V_T$, whereas for the depletion load mode, the voltage is close to V_{DD}. So in this case the difference in voltage between the logic 1 and logic 0 states is greater and the noise margins are also greater. This is a very important point and makes the use of depletion mode loads almost universal, the noise-margin advantage outweighing the disadvantage of the extra processing for the channel implant. A further advantage of the depletion mode option can be seen from the calculations above, in that the ratio of transistor sizes is smaller, hence the overall device size is potentially smaller.

3.5.3 Inverter transfer characteristics

From here on we shall only consider the depletion mode load inverter. The calculations above have only considered the output logic 1 and 0 voltages, that is the end points of the V_{out} against V_{in} transfer curve. Another important design consideration is the position of the transition curve, that is the value of V_{in} when V_{out} is half-way between its extreme values. Again, for best noise margins it is desirable to have this transfer curve anti-symmetrical, that is (assuming V_{out} swings between 0 V and V_{DD}) $V_{in} = V_{out} \approx V_{DD}/2$. At this point it can be assumed that both devices are in saturation (but see the detailed analysis below) and as both have the same drain current, we can equate these as follows:

$$\frac{W_d}{L_d}(V_{GSd} - V_{Td})^2 = \frac{W_\ell}{L_\ell}(V_{GS\ell} - V_{T\ell})^2$$

Using $V_{DD} = 5$ V, then $V_{GSd} = 2.5$ V, and using the values we had previously, we obtain

$$\frac{W_d}{L_d}(2.5 - 1)^2 = \frac{W_\ell}{L_\ell}(-3)^2$$

$$\frac{(W_d/L_d)}{(W_\ell/L_\ell)} = \frac{9}{2.25} = 4.0$$

This is very close to the previously derived value, so fortunately both design criteria can be fulfilled at the same time.

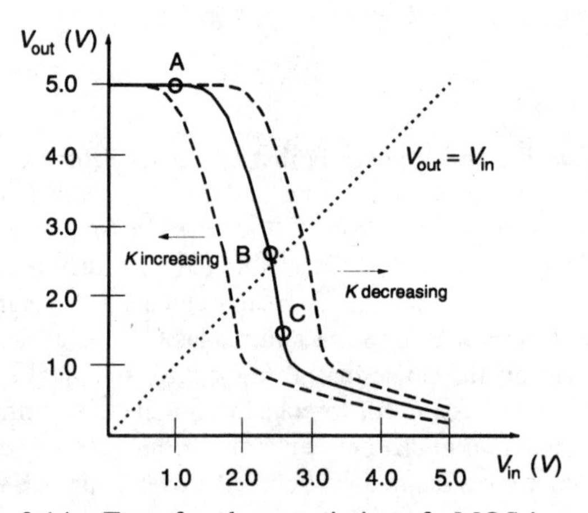

Figure 3.14 Transfer characteristics of nMOS inverter stage

Based on the ratio of transistor geometries of 4, as derived above, we can now examine the transfer characteristics in detail (see Figure 3.14). Starting with $V_{in} = 0$ V, the driver is cut off and the output voltage is very close to V_{DD}. This will continue to be the case until V_{in} rises above V_{Td} when the driver starts to conduct (point A). As V_{DSd} (V_{out}) exceeds $V_{GSd} - V_{Td}$ the driver is saturated. In addition $V_{DS\ell}$ ($V_{DD} - V_{out}$) is less than $|V_{T\ell}|$ so it is operating within the linear region acting as a resistor. As V_{in} increases, more current is drawn through the transistors, and V_{out} decreases. The line in this region is nearly linear, ideally, and with a constant value of $V_{T\ell}$ this would be the case. However, there is a significant source–substrate voltage (V_{out}) hence the non-linearity. This characteristic continues until $V_{out} = V_{DD} - |V_{T\ell}|$, which given the ideal case would occur at $V_{out} = 2$ V. In practice with the body effect the transition takes place at a lower V_{in} (point B); at this point $V_{DS\ell}$ exceeds $|V_{T\ell}|$ and the load transistor moves into its saturated region. Point B occurs approximately at the middle of the characteristic, and in deriving the optimum *W/L* ratios it was assumed that both transistors were in saturation. However, the exact point depends on the body effect and other factors that are too detailed to discuss here. The assumption made is one that is generally accepted, and the result of a ratio of 4 is in common use in the design of nMOS circuits.

The curve follows a different non-linear shape as V_{in} increases until a point is reached when $V_{DSd} = V_{out}$ is less than $V_{GSd} - V_{Td}$ (point C). At this point the driver transistor moves into the linear region. Using the relevant equations and balancing the current as before, this will occur when

$$\frac{W_d}{L_d}\left[(V_{in} - V_{Td})\,V_{out} - \frac{V_{out}^2}{2}\right] = \frac{W_\ell}{L_\ell}\frac{(-V_{T\ell})^2}{2}$$

At point C, $V_{out} = V_{in} - V_{Td}$ so we have

$$8\left[V_{out}^2 - \frac{V_{out}^2}{2}\right] = (3)^2$$

Therefore $V_{out} = 1.5$ V and $V_{in} = 2.5$ V, which is approximately the mid-voltage point, hence the very steep curve between points B and C. As the input voltage is increased to 5 V, V_{out} drops to its final voltage of 0.3 V (derived from the same equations above).

Also indicated in Figure 3.14 is the way in which the transfer characteristics depend on the ratio of gate geometries. The calculations were made based on

$$\frac{(W_d/L_d)}{(W_\ell/L_\ell)} = K = 4.$$

Broken curves are shown as this value is increased and decreased. The main effect of deviation due to non-ideal value of K is that the logic threshold value (when $V_{out} = V_{in}$) no longer occurs at $\approx V_{DD}/2$.

3.5.4 Inverter switching speeds

As can be seen from Figure 3.13, the nMOS inverter drives a capacitive load which consists of the gate capacitances of further transistors that the inverter drives, and also parasitic capacitances associated with the circuit interconnections. So when an inverter is driven by a rising or falling voltage edge, the output voltage will not be able to respond instantaneously, as there will be a certain time required to charge or discharge C_L. To estimate this effect we shall consider a very simple model of the charge or discharge process by a single *RC* network, the resistance being the channel of the device through which the charge current flows, and the capacitance being C_L. We have two cases, for rising and falling inputs as indicated in Figure 3.15.

For the rising edge, the output voltage is at logic 1 (5 V) initially, and after V_{in} has risen to V_{DD}, it falls to 0.3 V, the discharge current flowing through the

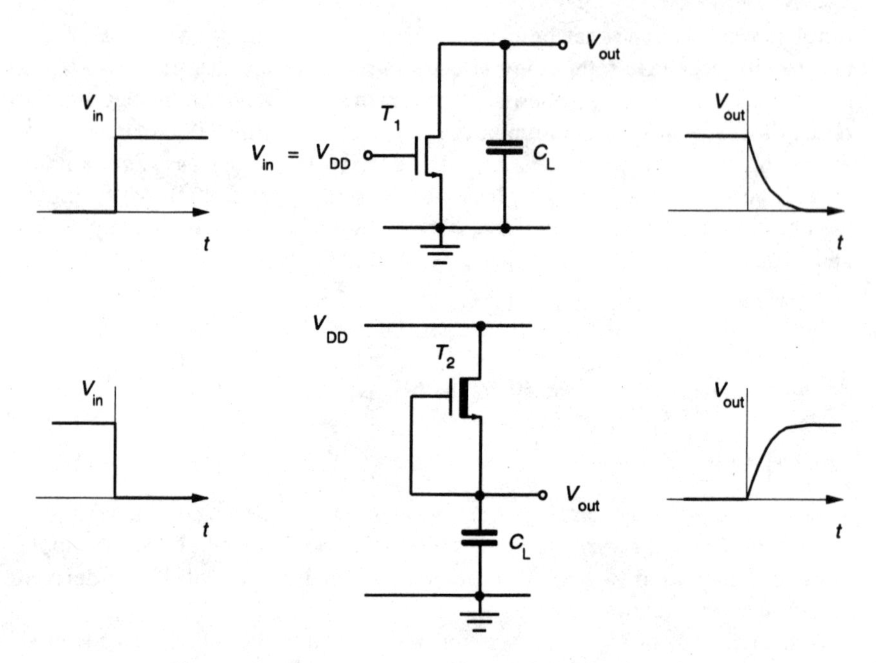

Figure 3.15 Edge transitions of nMOS inverter

driver transistor to ground. For the falling edge, the output voltage is at 0.3 V initially, and after V_{in} has fallen to logic 0, it rises to 5 V, the charge current flowing through the load transistor from the supply rail. The channel resistance depends on the carrier concentration, either induced or implanted, and also on the *W*/*L* ratio. The longer the channel, the greater the resistance of the device, hence charging times are directly proportional to *L*/*W*. As this ratio is generally four times lower for the driver transistor, T_1 will generally be shorter than T_2, unless there is a large difference in the channel conductivity, and this transition is the one that will provide the fan-out limit. As more gates are subsequently driven, or the interconnect lines extended too far, the charging time will become comparable with the clock period and errors can occur as logic transitions are not completed.

3.5.5 Other nMOS gates

Having considered the structure and operation of the nMOS inverter in some detail, we shall now look briefly at some other nMOS gates. The principal Boolean functions that are easily realizable in this technology are the NAND and NOR. These can be achieved by having a series of series or parallel enhancement mode devices connected to the inputs, as was mentioned briefly in Chapter 2 and illustrated in Figure 2.13. In order to keep the logic threshold value as close to the mid-rail voltage as possible, we must look again at the ratio of the transistor geometries for these gates.

In general we shall try to maintain the ratio of 4 for the driver *W*/*L* to load *W*/*L*, but this is complicated by the fact that there is more than one driver transistor. For the NAND device, if at least one of the inputs is logic 0, then there will be no current flow through the serial drivers; all the inputs must be logic 1 before the current flows and the logic transition takes place. Therefore the effective *W*/*L* for the drivers must be the sum of the individual driver *W*/*L* values. If, for example the *L*/*W* ratio for the load is 4, then the total *L*/*W* ratio for the drivers must be 1. If there are, say, two inputs, then the *L*/*W* ratio of each is $\frac{1}{2}$.

For the NOR gate, a minimum of one of the inputs has to be at logic 1 before a driver transistor conducts a current and the logic transition takes place. So the driver transistors have the same geometry ratio as if they were single inverter drivers to maintain the correct logic threshold value. If more than one input is at logic 1, then the total *K* value is different from the ideal, but as it is unlikely that more than one input voltage changes at exactly the same time, in terms of the logic threshold, this is likely to be affected by the change on a single driver transistor. The final voltage corresponding to the logic 0 output will depend on the number of driver transistors switched on, but as the output voltage need only be well below the V_T of subsequent transistors driven by the gate, the precise value is not critical. It is more important to have the correct logic threshold and therefore the optimal noise margins.

3.5.6 Pass transistors

As well as forming gate circuits, nMOS transistors can be used in their switched mode to control the passage of logic signals from one point in the circuit to another. They are operated as 'pass' transistors, as illustrated in Figure 3.16. With a logic 1 on the gate, the device is in a low-resistance state and another logic signal can pass from source to drain with only a little attenu-

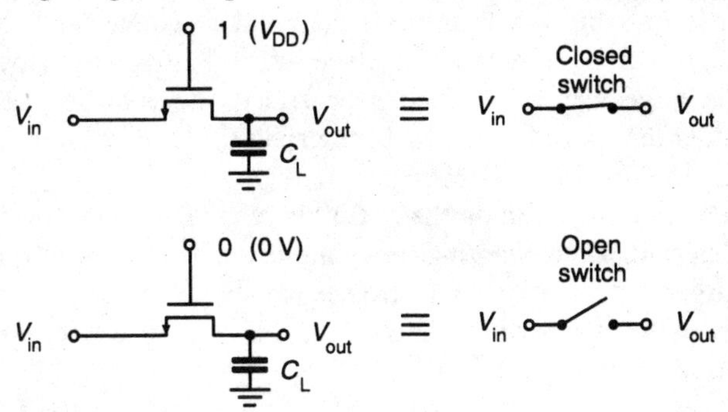

Figure 3.16 Operation of nMOS pass transistor

ation. In fact, the logic 0 (0 V) will be transferred exactly, and the logic 1 (5 V) will be reduced by the V_T of the pass transistor. With a logic 0 on the gate, the pass transistor is in a high-resistance state and the other logic signal cannot be transferred.

The on resistance of the pass transistor will have an effect on the rise time of the output voltage when charging a capacitive load. There are two design considerations here. The overall size of the device should be kept as small as possible, as with all IC elements, but the resistance must be low enough for the pass transistor to operate at the desired clock speed. A low-resistance channel means a short, wide device, but an increased device width means a larger area. In general it is found that a minimum size device (1:1 W:L) has a low enough resistance for normal operating speeds.

Owing to the reduced logic 1 value transferred by the pass transistor (typically 3.2 V), the geometry ratios of inverters or other gates driven by the pass transistor have to be adjusted to maintain the correct logic threshold. This normally requires a value of K of 8 instead of the usual 4. Alternatively a logic restoring inverter or buffer can be used after a pass transistor to restore the normal 5 V logic 1 level.

In some circuits, pass transistors are used in cascade as complex signal processing paths are constructed. The only major problem with this circuit arrangement is the delay effect. Each pass transistor provides a capacitive loading, C, to the previous stage, and in the on state each has a path resistance, R. So a cascade of n pass transistors can be modelled as a cascade of RC stages, and the delay associated with such a circuit increases as n^2. So for a particular circuit technology and operating speed, there is a natural limit to the length of such a cascade; typically $n = 4$ before the signal has to be buffered to restore the edges.

3.5.7 Buffer circuits

Certain signal lines on a chip will be required to drive a large capacitive load. Usually these are such signals as clocks that may be connected to the inputs of many gates, or output signals which have to drive off-chip loads. To be able to drive the larger capacitive loads, the signals have to be passed through buffer circuits that have increased drive capability, and these circuits can be inverting or non-inverting as required.

It was seen in the section on the nMOS inverter that the size of the capacitive load determines the charging time, and hence the speed of a logic transition. The time delay is proportional to the resistance of the transistor channel through which the charge or discharge takes place, and so also proportional to the L/W ratio, that is

$$\tau_d \propto \frac{C_L}{W/L}$$

Thus a bigger load capacitance could be driven by increasing the width of the transistors. However this increases the gate area and hence the capacitance that the previous stage drives: all we have done is to transfer the time delay problem to a previous stage of the circuit. The solution is to increase the drive capability of the stages gradually, as indicated in Figure 3.17. Here the W/L value is increased at each stage by a common factor n, and there are a total of m stages to drive the load C_L. Given a particular value of C_{in} and C_L there exists optimal values for n and m in terms of the total delay of the cascade. We have seen that having a single stage does not improve the delay problem. At the other extreme, having lots of stages with a small increase in W/L would also result in a large delay from the sum of the individual stage delays.

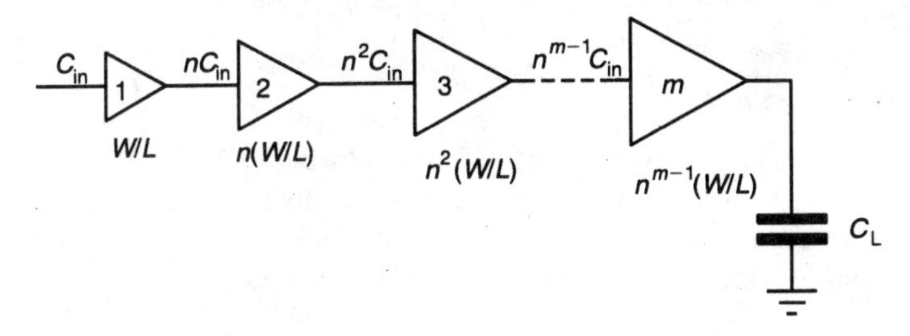

Figure 3.17 Cascade of buffers to drive large capacitance load

First we derive an expression for the number of stages m. The values of n and m are such that

$$C_L = n^m C_{in}$$

Rearranging

$$m = \frac{\ln(C_L/C_{in})}{\ln(n)}$$

Now considering the delay associated with the ith stage, this can be estimated by the ratio of the load capacitance seen by the stage to the W/L value (as above). The capacitance is given by $n^{i-1}(nC_{in})$ and the gate ratio by $n^{i-1}(W/L)$. So the delay for each stage is

$$\tau_i \propto \frac{n^{i-1}(nC_{in})}{n^{i-1}(W/L)}$$

$$\tau_i \propto \frac{nC_{in}}{W/L}$$

The right-hand side is independent of i and hence the delay associated with each stage is the same. The total delay of the cascade is therefore

$$\tau_d \propto \frac{nC_{in}m}{W/L}$$

Substituting in for m we get

$$\tau_d \propto \frac{nC_{in}\ln(C_L/C_{in})}{(W/L)\ln(n)}$$

In terms of the variation of τ_d with n

$$\tau_d \propto \frac{n}{\ln(n)}$$

To find the optimal value of n, we differentiate this expression with respect to n and set this equal to zero to find the minimum value of τ_d

$$\frac{d\tau_d}{dn} = \frac{\ln(n) - 1}{\ln(n)^2}$$

This expression is zero when $\ln(n) = 1$, that is when $n = e$ (2.72). Hence this is the value by which the size of the transistor geometries should be increased to minimize the overall delay of the cascade. The number of stages required depends on the value of the capacitive load to be driven

$$m = \ln(C_L/C_{in})$$

Therefore, for example, if the ratio of the capacitances is 1000, seven stages would be required. In practice, when very large load capacitances need to be driven, the area cost of having many stages becomes a dominant factor, and it is often better to have fewer stages, increase the value of n and suffer an increased total delay time. This increase will only be slight because the function $n/\ln(n)$ only varies slightly for values of n between 2 and 6.

Another disadvantage of the standard nMOS inverter is the fact that the delay associated with the two logic transitions is different, as described in an earlier section. This can especially be a disadvantage when the inverter is driving a load capacitive load, as the longer transition delay can severely limit the operating speed of the circuit. A solution to this problem is the so-called superbuffer which can be in an inverting or a non-inverting form, as illustrated in Figure 3.18.

The operating advantage of the superbuffer can be easily explained by analysing the operation of the inverting circuit. We shall assume that we have the normal K factor of 4 for the transistor geometries. The positive going input of V_{in} rising from 0 V to 5 V follows the behaviour seen in the standard inverter. T_1 and T_3 are both turned on. The inverter circuit of T_1 and T_2 is loaded by the gate of T_4, which has a capacitance four times that of the usual input gate,

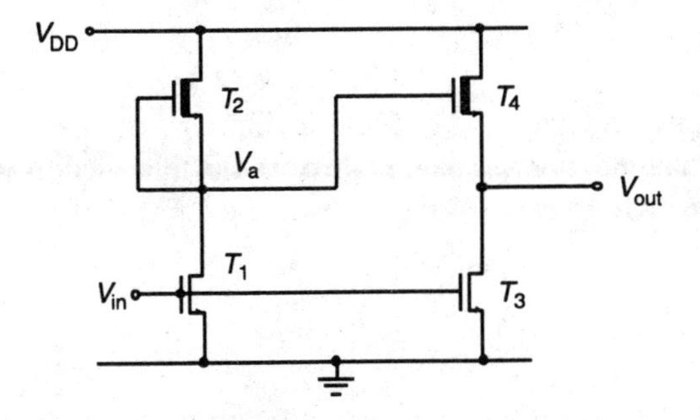

Inverting superbuffer

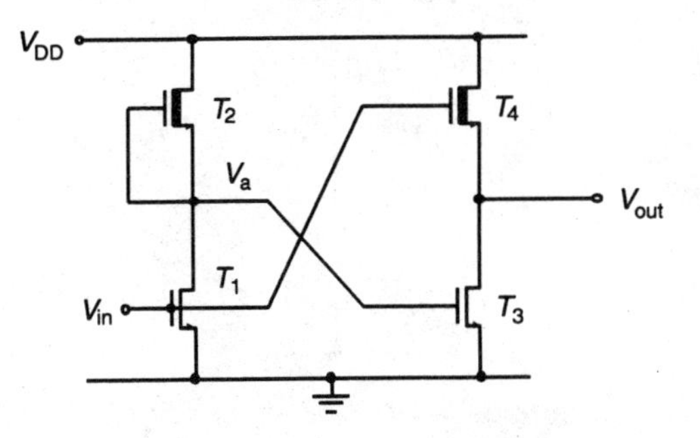

Non-inverting superbuffer

Figure 3.18 Superbuffer circuits

V_a falls to 0 V with about four times the normal delay (that is, this inverter stage has an effective fan-out of 4). T_4 becomes cut off and V_{out} falls to 0 V with about the same time delay.

Now consider the negative going edge as V_{in} changes from 5 V to 0 V. This is the transition that in the normal inverter has a longer associated delay, by a factor of about K. Again the inverter of T_1 and T_2 is loaded by the gate of T_4. For the inverter of T_3 and T_4, V_{in} is also applied to the gate of T_3, but the gate of T_4 is supplied by V_a. As the transition takes place and V_a rises towards V_{DD} the gate–source voltage of T_4 is greater than it would be if it were strapped to the source as in the normal inverter configuration. As the drain current is proportional to the square of V_{GS}, the current driving capability is increased and the charging delay reduced by the same factor. On average the current capability is increased by about a factor of 4 and the transition delay reduced by the same factor, bringing it to about the same value as the positive going delay and thus equalizing the delays of the two transitions.

3.6 CMOS gates

As was described in Chapter 2, CMOS technology consists of pairs of complementary n-channel and p-channel devices. In this section we shall look at the design considerations of digital gates formed in this way.

3.6.1 The CMOS inverter

The basic CMOS inverter is illustrated in Figure 3.19, while its transfer characteristic is shown in Figure 3.20. The operation of the inverter circuit is rel-

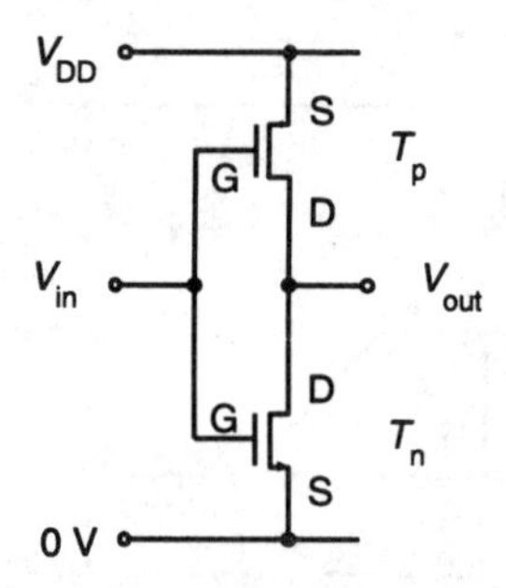

Figure 3.19 CMOS inverter circuit

atively straightforward as are the design considerations for the logic threshold. Both devices are enhancement mode and we shall assume that they both have threshold voltages of 1 V. In this case the mobility of the dominant charge carriers is different, as the charge carriers are electrons in the n-channel device and holes in the p-channel device. The mobility of holes is less than that of electrons, and this difference must be compensated for in the transistor geome-

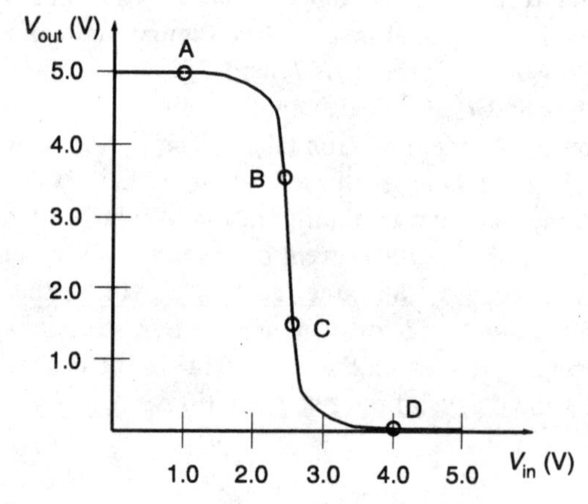

Figure 3.20 Transfer characteristics of CMOS inverter stage

tries to ensure that the correct logic threshold and a symmetrical transfer characteristic are achieved.

We shall now examine the transfer characteristic in detail. With $V_{in} = 0$ V, V_{GSn} is below V_T and this device is cut off. V_{GSp} is 5 V, so this device is on, V_{out} is about 5 V and V_{DSp} is about 0 V, so T_p is in its linear region. In this state only a very tiny leakage current will flow through the two transistors.

As V_{in} is increased beyond 1 V (point A), T_n starts to conduct, and V_{out} starts to fall. V_{DSn} is greater than $V_{GSn} - V_T$, so T_n is now in saturation. Initially T_p remains in its linear region as V_{DSp} is still small. Increasing V_{in} results in a non-linear drop in V_{out} and an increase in the current flow through the transistor pair. This continues up to the point where $V_{DSp} = V_{GSp} - V_T$ (point B). At this point T_p goes into saturation. V_{out} drops rapidly with increasing V_{in} until $V_{DSn} = V_{GSn} - V_T$ when T_n will transfer into the linear region (point C). Increasing V_{in} to beyond 4 V (point D) results in V_{GSp} being less than V_T and so T_p is cut off, virtually no current flows and the V_{out} is approximately 0 V.

For the best noise margins we wish the logic threshold to be at $V_{DD}/2$. At this centre point both devices are in for saturation, the currents are given by

$$I_{Dn} = \frac{W_n \varepsilon \mu_e}{2L_n D} (V_{GSn} - V_T)^2$$

$$I_{Sp} = \frac{W_p \varepsilon \mu_h}{2L_p D} (V_{SGp} - V_T)^2$$

These currents are identical as the two transistors are in series, so for the threshold condition, $V_{GSn} = V_{SGp} = 2.5$ V and we have

$$\frac{W_n/L_n}{W_p/L_p} = \frac{\mu_h}{\mu_e}$$

and the value of K is determined by the ratio of the mobilities. In silicon μ_e is about twice the value of μ_h, so a K value of about 2 is often used.

Returning now to the transfer curve, we have seen that point A occurs when $V_{in} = 1$ V, $V_{out} = 5$ V, and point D when $V_{in} = 4$ V, $V_{out} = 0$ V. In the analysis of the central part of the transfer characteristics (between points B and C) to determine the K factor, it can be seen that V_{out} does not appear in the equations, which indicates that this section is independent of V_{out} and hence the line is vertical in this section. In fact this is an approximation and channel modulation effects will cause a finite but very steep slope. We shall use this approximation to estimate the values of V_{out} for points B and C.

For point C, T_p is in saturation and T_n just in the linear region. Equating currents for these conditions and with the correct K factor

$$\frac{(V_{SGp} - V_T)^2}{2} = (V_{GSn} - V_T)\, V_{DSn} - \frac{V_{DSn}^{\ 2}}{2}$$

and in terms of V_{in} and V_{out}

$$\frac{(V_{in} - V_T)^2}{2} = (V_{in} - V_T)\, V_{out} - \frac{V_{out}^{\,2}}{2}$$

With $V_{DD} = 5$ V, $V_{in} = 2.5$ V and $V_T = 1$ V, this gives a quadratic in V_{out}:

$$V_{out}^{\,2} - 3V_{out} + 2.25 = 0$$

which gives $V_{out} = 1.5$ V. Owing to symmetry, for point B, $V_{out} = 3.5$ V.

As the asymmetry in the transistor geometries is essentially only to compensate for the difference in the carrier mobilities, both devices have the same driving capability and hence the positive and negative going transitions are virtually symmetrical.

3.6.2 Other CMOS gates

NAND and NOR gates are easily implemented in CMOS technology and these two gates are illustrated in Figure 3.21. These are two-input circuits, but the number of inputs can be easily increased by keeping the same topology.

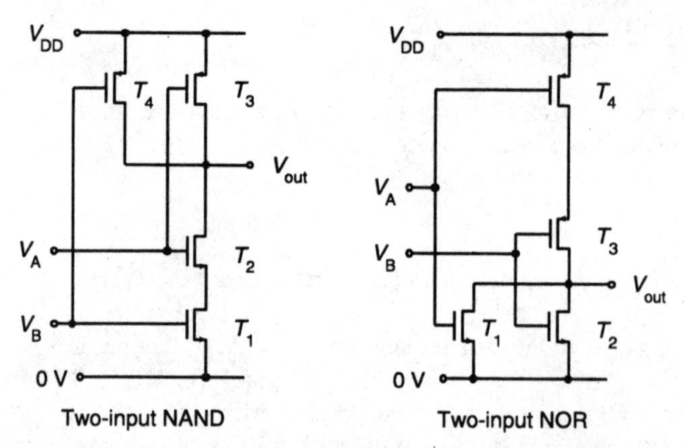

Figure 3.21 CMOS NAND and NOR circuits

The operation of the two gates is straightforward. For the NAND gate, if either or both of the input voltages are below the V_T of the n-channel devices then at least one of T_1 and T_2 is cut off, while at the same time at least one of T_3 and T_4 is turned on. V_{out} will go to V_{DD}, the charging current coming through T_3 and/or T_4. When both inputs are at V_{DD}, then both T_1 and T_2 are conducting, while both T_3 and T_4 are cut off, so V_{out} goes to 0 V, this time the charging current being supplied through T_1 and T_2 in series.

The operation of the NOR gate is similarly explained. When either or both of V_A and V_B are equal to V_{DD}, at least one of T_1 and T_2 is conducting and at least one of T_3 and T_4 is cut off so V_{out} goes to 0 V, the charging current being passed through T_1 and/or T_2. When both inputs are at 0 V, both T_1 and T_2 are cut off and both T_3 and T_4 are conducting. V_{out} goes to V_{DD} with the charging current passing through T_3 and T_4 in series.

The problem with these CMOS gates is that the symmetry of edge transition times cannot be guaranteed; the particular charging resistance will depend on the input logic combination as to which transistors are conducting and which are cut off. Some adjustment of the relative transistor geometries can be made to offset the worst-case asymmetry, but the asymmetry cannot be entirely eliminated. It is more common for designers to use minimum-size devices to reduce circuit area and simply live with the consequences of the delay asymmetry.

3.6.3 Transmission gate

The CMOS equivalent of the nMOS pass transistor is the so-called transmission gate, as illustrated in Figure 3.22. As MOS devices are usually symmet-

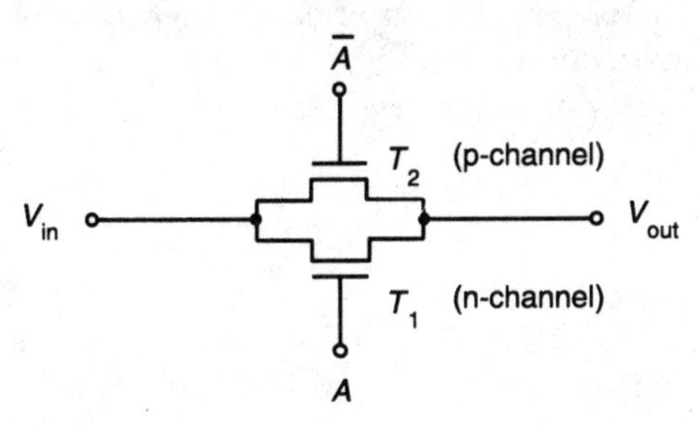

Figure 3.22 CMOS transmission gate

rical, the source and drain terminals are effectively interchangeable, depending on the voltage levels, that is the drain is usually at a higher positive voltage than the source in the n-channel device. As either side of the transmission gate can be at a higher voltage, the terminals of the devices in the gate can interchange, hence in Figure 3.22 there are no arrows drawn on the transistors to indicate their orientation.

When the signal A is at logic 0 then both of the devices are turned off, only a very small leakage current can flow through the gate and V_{out} remains as it was. If the signal A is brought high, then there is V_{DD} on the gate of the n-channel device and 0 V on the gate of the p-channel device, and both are turned on. Current can now flow through the gate, depending on the relative values of V_{in} and V_{out}. Clearly if both voltages are at the same level, then again no current will flow through the gate. If the voltages are at different logic levels, there are two cases to consider:

(a) V_{in} at logic 1 and V_{out} at logic 0. In this case V_{in} is connected to the drain of T_1 and the source of T_2. Current flows through the gate from V_{in} to V_{out}, charging any load capacitance. During this process V_{SGp} remains at V_{DD}, so T_2 remains turned on throughout. V_{GSn} starts at V_{DD}, but as the capacitor charges up, this gate–source voltage will fall to 0. When it drops below V_{Tn}, T_1 will cut off and the charging can only continue through T_2.

(b) V_{in} at logic 0 and V_{out} at logic 1. Here V_{in} is connected to the source of T_1 and the drain of T_2. Current flows through the gate from V_{out} to V_{in} discharging the load capacitance. In this case V_{GSn} remains at V_{DD} throughout and T_1 is turned on throughout. V_{SGp} starts at V_{DD} but drops as the load capacitance discharges. When V_{out} drops below V_{Tp}, T_2 cuts off and the discharge current can only flow through T_1.

So in both cases the charging process takes place in two stages: a fast charge through the two parallel devices, followed by a slower charge rate through a single device. However, unlike the nMOS pass transistor, the output voltage will charge to the full value of V_{DD}, but the transmission gate suffers the disadvantages of an increased area and the requirement for complementary signals to be available, or the addition of an inverter to derive the extra signal. The ratio of the transistor geometries can be made the same as in the CMOS inverter to provide a symmetrical operation.

3.7 Analogue circuits

Like digital circuits, analogue ICs and systems can be broken down into small building blocks, but the function of the analogue blocks tends to be more diverse than that of digital gates, which largely just implement simple Boolean functions. This is due largely to the nature of analogue circuits which operate on a continuum of signals, both in magnitude and frequency (or time). In practice, digital circuits can be considered as a special case of analogue circuits, as has been seen in the previous sections in this chapter, where the behaviour of the digital gates has been analysed on a voltage continuous or analogue basis.

We shall now look at some of the details of these building blocks, and the main design rules associated with them.

3.7.1 Capacitors

Along with the transistor elements, which have been dealt with in Chapter 2, the capacitor is the most important circuit element in analogue ICs. It finds little use in digital circuits, but has a number of roles in analogue circuits, for example in switched capacitor circuits, as compensation elements in amplifiers, as filter elements and as reference and charge splitting components in analogue-to-digital converters (ADCs) and digital-to-analogue converters (DACs). The important point about monolithic capacitors is that although the absolute value of capacitance cannot be guaranteed, owing to processing variations, the ratio of component values can be very precise, and much of the circuit design is based around this principle.

Monolithic capacitors are very easy to produce, being almost universally based on the parallel plate system in the planar process. The dielectric material is most often silicon dioxide, which is easily deposited and forms a very stable dielectric. Silicon nitride is also used, providing a higher dielectric constant, and therefore higher capacitance per unit area. The conducting plates may be formed using the metallization layer(s), the conducting polysilicon layers, or the diffusion areas. The preferred system is to use polysilicon for both layers, but this is only available in the more complex MOS technologies. In bipolar technologies, the most common choice is a metal-diffusion device.

As well as the basic parallel plate capacitance value associated with the area of the overlapping plates, there are always parasitic fringing capacitances present, which can more than double the total capacitance and which are associated with the length of the periphery of the device. These fringing capacitances also depend on which layers comprise the conducting plates and the proximity of any other layers. The difficulty in forming a capacitor of a precise value can thus be appreciated. As was mentioned earlier, many designs are based on capacitance ratios between different devices, so one approach to ensure exact ratios is to use many small capacitors ('unit capacitors') and then connect a number of these in parallel to form larger values of capacitance.

3.7.2 Switches

Active switches are becoming more and more common in analogue circuit realizations of such circuits as ADCs and DACs, as well as, of course, switched capacitor circuits. The realization of analogue switches is identical in form to that of digital pass transistors and transmission gates, so we will not repeat the analysis of those components in this section. Only MOS switches were described earlier, so it is worth noting that bipolar devices can be used as switches in a similar configuration to the MOS pass transistor. The base has the switching signal applied to it, and the active signal is passed through from emitter to collector. The bipolar switch is not as good as the MOS device – the characteristics are more non-linear, the finite base current means that the signal current flowing through the switch changes, and there is an offset voltage associated with the on state.

Switches in analogue circuits are usually controlled by clock waveforms in a regular timing pattern. There may be a number of overlapping or non-overlapping clock signals in the system, depending on the complexity of the circuit function. One mechanism that arises and can cause problems in some circuits is clock feedthrough in the switches. This occurs because of the presence of parasitic capacitances in the MOS transistor, which connect the gate to the source and drain regions. Although in most devices these capacitances are very small, of the order of femtofarads, with the large signal, fast switching of the clock signal on the gate, a significant voltage can be induced on the transmitted signal. This can be a particular problem when the level of the transmitted signal is low, but needs to be accurately passed, for example in highly accurate ADCs. There is little that can be done to eliminate clock feedthrough in nMOS circuits. Having minimum geometry switches and keeping clock voltage swings small can reduce the effects, but these restrictions may cause other problems in the overall design. CMOS transmission gates do not suffer so badly from clock feedthrough as the two transistors are driven by oppositely phased clock signals, so the feedthrough from the two devices tends to cancel out. Transmission gates have an additional advantage of higher dynamic range, as the output signal can reach the full supply rail voltage, as described in sections 3.5.6 and 3.6.3.

3.7.3 Reference circuits

As will be seen in later sections, there is often a requirement for a sub-circuit that will provide an accurate reference value of voltage (or sometimes current). These circuits should be very stable, ideally having zero or very well known temperature variations, and do not suffer variations as a result of other external influences, such as power supply variation.

The simplest form of voltage reference circuit employs a single transistor, and both the bipolar and MOS versions are illustrated in Figure 3.23. Here the transistors are acting as active resistors and form a simple potential divider circuit with the resistance *R*; the equivalent circuit is also illustrated in Figure 3.23. The reference voltage is therefore given by

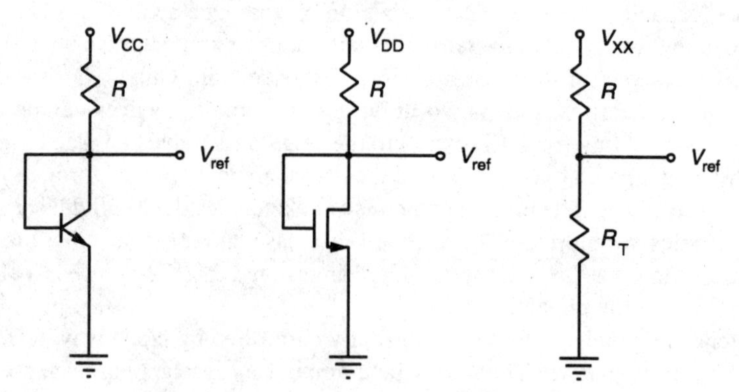

Figure 3.23 Simple voltage reference circuits

$$V_{\text{ref}} = V_{\text{XX}}\left(\frac{R_{\text{T}}}{R + R_{\text{T}}}\right)$$

This of course assumes no significant loading effect from the circuit V_{ref} supplies. We shall be concerned primarily with the effect on V_{ref} of variations in temperature and supply voltage, V_{XX}. We can define a fractional temperature coefficient (TC) and a supply current sensitivity (S) as follows:

$$\text{TC}(V_{\text{ref}}) = \frac{1}{V_{\text{ref}}}\frac{\partial V_{\text{ref}}}{\partial T}$$

$$\text{S}(V_{\text{XX}}) = \frac{V_{\text{XX}}}{V_{\text{ref}}}\frac{\partial V_{\text{ref}}}{\partial V_{\text{XX}}}$$

So for the equivalent circuit in Figure 3.23 we have

$$\begin{aligned}
\text{TC}(V_{\text{ref}}) &= \frac{1}{V_{\text{ref}}}\frac{\partial V_{\text{ref}}}{\partial T} \\
&= \frac{R+R_{\text{T}}}{R_{\text{T}}\,V_{\text{XX}}}\frac{\partial}{\partial T}\left(\frac{V_{\text{XX}}R_{\text{T}}}{R + R_{\text{T}}}\right) \\
&= \frac{R+R_{\text{T}}}{R_{\text{T}}}\frac{\partial}{\partial T}\left(\frac{R_{\text{T}}}{R + R_{\text{T}}}\right) \\
&= \frac{R+R_{\text{T}}}{R_{\text{T}}(R+R_{\text{T}})^2}\left[\frac{\partial R_{\text{T}}}{\partial T}(R+R_{\text{T}}) - \left(\frac{\partial R}{\partial T} + \frac{\partial R_{\text{T}}}{\partial T}\right)R_{\text{T}}\right] \\
&= \frac{1}{R_{\text{T}}(R+R_{\text{T}})}\left[R\,\frac{\partial R_{\text{T}}}{\partial T} - R_{\text{T}}\,\frac{\partial R}{\partial T}\right] \\
&= \frac{R\,R_{\text{T}}}{R_{\text{T}}(R+R_{\text{T}})}\left[\frac{1}{R_{\text{T}}}\frac{\partial R_{\text{T}}}{\partial T} - \frac{1}{R}\frac{\partial R}{\partial T}\right] \\
&= \frac{R}{(R+R_{\text{T}})}\,[\text{TC}(R_{\text{T}}) - \text{TC}(R)]
\end{aligned}$$

If the temperature coefficients of the two resistances are the same, then the temperature coefficient of V_{ref} will be zero. If two monolithic resistances are formed from the same layer, for example, by diffusion, they will have the same TC and $\text{TC}(V_{\text{ref}})$ is zero. For the pure resistor implementation, to avoid loading effects and to keep the size of the components small, the resistors would have to have a small value. This would result in very large current consumption and power wastage, hence the use of active resistors. The problem here is matching of the TC of the active device to the monolithic resistor. The temperature behaviour of the transistors is very complex, but it is found that in general the TC(V_{ref}) for the bipolar circuit is fairly poor, that of the MOS circuit is better, but improved circuits are available as will be described shortly.

Before we move on, we should also look at the sensitivity of these circuits to power supply variation. This was given before as

$$S(V_{XX}) = \frac{V_{XX}}{V_{ref}} \frac{\partial V_{ref}}{\partial V_{XX}}$$

with

$$V_{ref} = V_{XX} \left(\frac{R_T}{R + R_T} \right)$$

which gives $S(V_{XX}) = 1$. This means that there is a direct relationship between variations in V_{XX} and V_{ref}. If the supply voltage varies by 10 per cent then the reference voltage will also change by 10 per cent. Ideally we need this sensitivity to be reduced to zero.

Current reference circuits also play an important part in analogue circuit design, and a simple implementation is shown in Figure 3.24, again with both

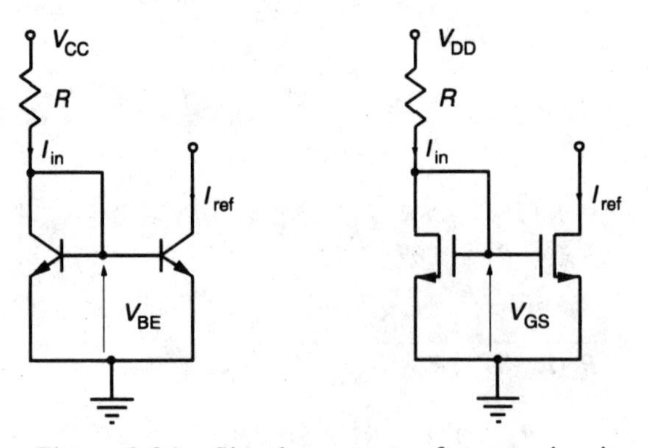

Figure 3.24 Simple current reference circuits

bipolar and MOS versions shown. These circuits are basic current mirrors; as the V_{BE} and V_{GS} are the same for each pair of transistors, then I_{ref} must be the same as I_{in}. It can also be seen that the left-hand side of each circuit is identical to the voltage reference circuits just described. In other words we are setting V_{BE} or V_{GS} equal to V_{ref} which then determines I_{ref}. Detailed analysis of the TC behaviour (see Geiger *et al.*, 1990, Chapter 5) shows that it is possible to obtain a zero value of TC by correct choice of V_{XX}. The supply voltage sensitivity of I_{ref} is still unity, however, as in the voltage reference circuit.

An improved voltage reference circuit, potentially independent of temperature and supply voltage variation, is the bandgap voltage reference. The basis of the circuit is to match the temperature variation of a p–n junction potential, which has a negative TC, with a thermally generated voltage reference

($\propto kT/e$), which clearly has a positive TC. By scaling and summing the two voltages, we can derive a reference voltage with a zero TC. If we use the base–emitter junction voltage V_{BE} of a BJT as the p–n junction potential and $V_t = kT/e$, we have

$$V_{ref} = V_{BE} + GV_t$$

where G is the required scaling factor, that is the ratio of the TC magnitudes.

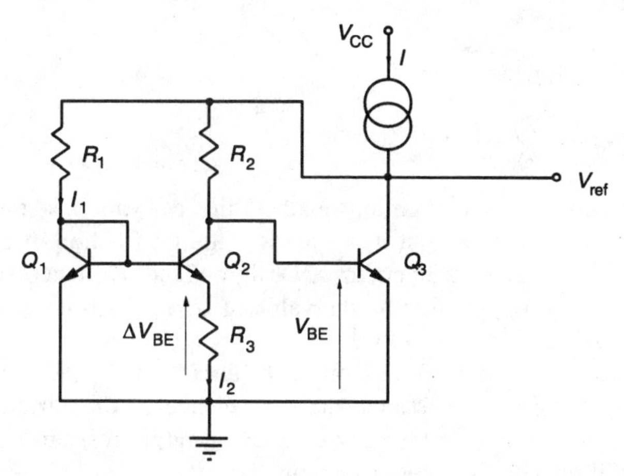

Figure 3.25 Widlar voltage reference circuit

A common form of the bandgap reference voltage is due to Widlar (1971) and in its simplest form is illustrated in Figure 3.25. Here we will assume that all the BJTs are identical in performance and that they have relatively high current gains, β. The presence of R_3 means that there is a difference in V_{BE1} and V_{BE2}, ΔV_{BE}, which we saw in section 3.4 was given by

$$\Delta V_{BE} = V_t \ln \frac{I_1}{I_2}$$

Now considering the base–emitter voltage of the BJT, and in particular its temperature dependence, this is given approximately by

$$V_{BE} = V_{g0}\left(1 - \frac{T}{T_0}\right) + V_{BE0}\left(\frac{T}{T_0}\right)$$

where V_{g0} is the extrapolated energy bandgap voltage for the semiconductor material at absolute zero of temperature (1.205 V for silicon) and V_{BE0} is the base–emitter voltage at temperature T_0.

If the current gain is sufficiently high, the current through R_2 is approximately the same as through R_3, so the voltage across R_2 is given by $(R_2/R_3)\Delta V_{BE}$ and the overall reference voltage given by

$$V_{ref} = V_{g0}\left(1 - \frac{T}{T_0}\right) + V_{BE0}\left(\frac{T}{T_0}\right) + \frac{R_2}{R_3} V_t \ln \frac{I_1}{I_2}$$

Comparing with the previous equation for V_{ref}, this gives the gain factor G to be $(R_2/R_3)\ln(I_1/I_2)$.

Now for zero temperature drift, we need to differentiate the above equation with respect to temperature, giving

$$\frac{\partial V_{ref}}{\partial T} = -\frac{V_{g0}}{T_0} + \frac{V_{BE0}}{T_0} + \frac{R_2}{R_3}\frac{k}{e}\ln\frac{I_1}{I_2}$$

which is zero when

$$V_{g0} = V_{BE0} + \frac{R_2}{R_3}\frac{kT_0}{e}\ln\frac{I_1}{I_2}$$

There are a number of approximations made in this derivation, so the temperature stability is not perfect, but it is usually within a fraction of 1 per cent over a very wide range of temperatures. Design of the resistance values can control the value of V_{ref} by altering the value of G, and also maintaining the temperature stability condition above.

Improvements in the basic Widlar reference circuit can be made by the use of operational amplifiers to remove the dependence of the currents on the value of the power supply and to provide a lower output resistance reference supply that will be more tolerant of loading.

3.7.4 Operational amplifiers

Operational amplifiers (op-amps) have become the most fundamental and useful building block in analogue ICs. Ideally their performance is that of a voltage-controlled voltage source with infinite gain, infinite input impedance and zero output impedance. Of course, practical circuits cannot have these characteristics, and there are a number of non-idealities in the response that may have to be taken into account in a practical design. These are discussed in the following paragraphs.

Finite gain and frequency response

The op-amp has a differential input configuration, with a non-inverting input (v^+) and an inverting input (v^-), such that the output voltage is given by

$$v_o = A_v\,(v^+ - v^-)$$

where A_v is the differential voltage gain, which, as noted above, is ideally infinite. In reality, very large values of A_v are possible, up to 10^6. However this value is very frequency dependent, and can be modelled as a first-order response as shown in Figure 3.26. An important measure of the op-amp, which takes both the d.c. gain value and the frequency response into account, is the gain–bandwidth product which is simply $A_v \times f_1$ where f_1 is the frequency at which the voltage gain has dropped to unity.

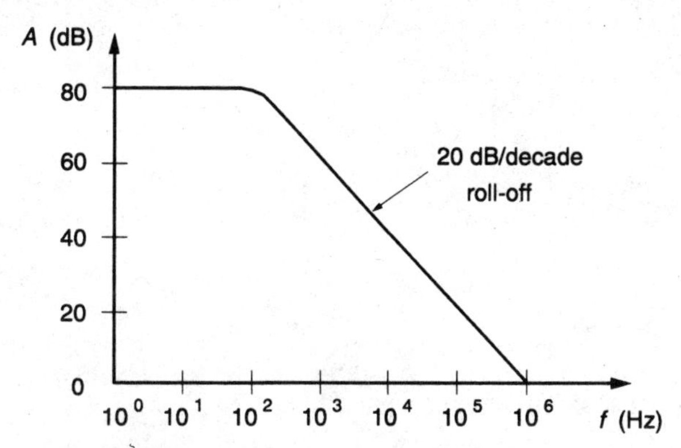

Figure 3.26 Typical op-amp gain–frequency curve

Linear range

With the large value of gain, it only takes a small differential voltage to generate a large output voltage. The output voltage cannot surpass the voltage of the power supply rails, and the linear relationship for v_o only holds for a range a little smaller than these limits.

Offset voltage

Ideally, when $v^+ = v^-$ the output voltage is zero, but in real devices it is a finite value, $v_{o,off}$. This is usually directly proportional to the gain, so can be thought of as an input offset voltage $v_{i,off}$, which is typically in the range 5–15 mV.

Common mode rejection ratio (CMRR)

We can define a common mode input voltage, $v_{i,c} = (v^+ + v^-)/2$, and with this voltage applied to both inputs and measuring the output voltage, v_o, the common mode gain is defined as $A_c = v_o / v_{i,c}$. This ideally should be zero, as above. The amount of non-ideality is defined relative to the normal gain with a differential input $(v^+ - v^-)$, A_d (equivalent to A_v above). This measure is known as the common mode rejection ratio and is defined as

$$\text{CMRR} = 20\log_{10}\frac{A_d}{A_c}$$

Typically this is in the range 60–80 dB, and describes how well the amplifier can suppress any common mode signals, such as noise.

Power supply rejection ratio (PSRR)

A similar effect to CMRR is the response of the amplifier to power supply variations. A small variation in the power supply voltage, v_p will result in a change in the output voltage by $A_p v_p$. The PSRR is also given in logarithmic units by

$$\text{PSRR} = 20\log_{10}\frac{A_d}{A_p}$$

PSRRs are of the same orders of magnitude as CMRRs.

Non-ideal input and output impedances

The input impedance of practical op-amps is not infinity but can be made very high, of the order of MΩ or higher, and for most practical purposes this is close to ideal. The output impedance can vary greatly from the ideal of zero, up to kΩ or even MΩ. It is usually resistive in nature, and the main effect is to slow the charging speed when feeding a capacitive load, and hence the frequency response of the overall circuit.

Slew rate

When the input voltage changes rapidly, the output voltage from the op-amp cannot always follow the change so quickly, but takes a longer time to reach its final value. A particular op-amp has a maximum rate of change of its output voltage that is termed the slew rate. It is usually quoted in terms of V μs^{-1}, and typical values are 1–10 V μs^{-1}.

Op-amps can be formed in BJT, nMOS or CMOS technologies and have similar structure and behaviour. The input and output resistances of MOS devices tend to be higher than BJTs, which usually have higher gain and lower offset voltages. Because of the similarities in circuit operation, just the details of the CMOS device will be given, as this technology is becoming more and more favoured by designers.

The block structure of a CMOS op-amp is shown in Figure 3.27. The input is a differential amplifier stage which provides a large amount of the amplification. As the other stages are single-ended in nature, the next block is a differential to single-ended converter stage. This is followed by a level shifter to compensate for the d.c. voltage change in the input stage and so provide the correct d.c. level for the following stages. The differential input stage cannot supply enough overall gain, so a further amplification stage is added, followed by an output buffer which is a unity gain stage with enough current drive and low output resistance so that the overall amplifier can drive large capacitive or resistive loads.

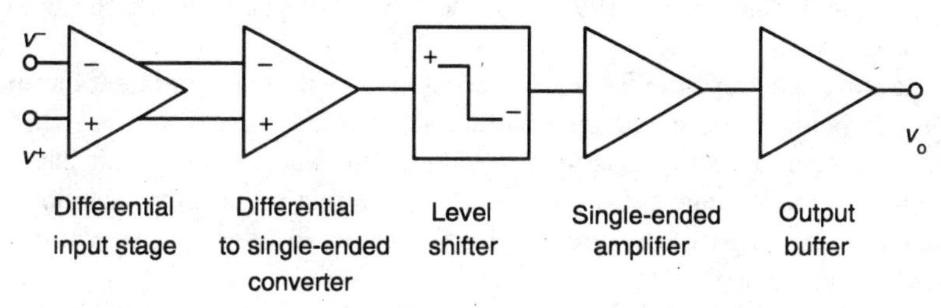

Figure 3.27 Block diagram of CMOS op-amp circuit

A simple CMOS op-amp circuit based on this topology is shown in Figure 3.28. In this circuit transistors T_1 to T_4 form the differential amplifier stage with T_5 providing a constant current source driven by the bias voltage. This voltage could be derived from a reference stage such as a Widlar circuit. The differential stage is such that it incorporates the single-ended output which is connected to the gate of T_6. This device acts as a driver which, in combination with T_7 as a load, forms the second-stage amplifier and also acts as the level

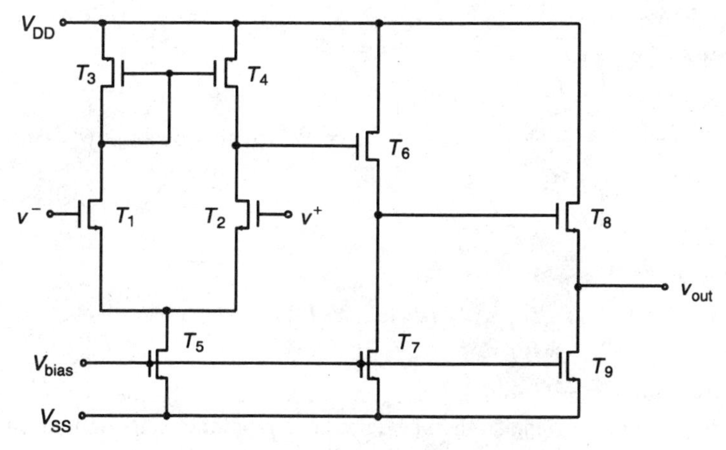

Figure 3.28 Basic CMOS op-amp circuit

shifter. The signal is then passed to T_8 which acts as a unity gain source, followed with T_9 acting as the active load.

The overall performance of the op-amp will depend on the geometries of the transistors, as we have seen in other MOS building blocks (this is different from the BJT case, where the performance is largely independent of the transistor geometries). This gives the designer extra degrees of freedom in the design of MOS circuits, although it does complicate the overall design somewhat. It is possible to specify a number of characteristics, such as gain, slew rate and gain–bandwidth product, and simultaneously satisfy them by correct choice of transistor geometries. Details of these designs can be found in other texts (such as Geiger *et al.*, 1990), and here we will just indicate the broad design approach.

The three-stage amplifier can be represented by a simple equivalent circuit, each stage being represented by a voltage-controlled current source with a load of a capacitor and a resistor in parallel. This equivalent circuit can be solved for overall voltage gain and, based on the effect of the capacitive loads, the frequency response of the circuit. This frequency response is a very important factor and we shall return to it shortly.

In the differential amplifier circuit, we want to keep the inverting and non-inverting behaviour as symmetrical as possible, so the sizes of T_1 and T_2 are the same, and likewise for T_3 and T_4. In terms of the d.c. conditions, this ensures that the currents through the two arms of the differential stage are the same and equal to half the current through T_5. The other biasing condition we need is for linear behaviour; the transistors must be operated in saturation. Most of the devices will be in saturation owing to the circuit connections or the applied voltages, but we must design the circuit to maintain T_4 in saturation. We can do this by ensuring that the V_{GS} of T_6 is the same as that of T_3. Then $V_{DS} = V_{GS}$ for T_4 and the saturation condition always applies. The current flow through parallel MOS devices is in direct proportion to their W/L values, so we have

$$\frac{W_3}{L_3} = \frac{W_6}{L_6}\left(\frac{I_3}{I_6}\right)$$

We also have $I_6 = I_7$, $I_3 = I_4$ and $W_3/L_3 = W_4/L_4$, so we can now write

$$\frac{W_4}{L_4} = \frac{W_6}{L_6}\left(\frac{I_4}{I_7}\right)$$

I_4 is half of I_5 so

$$\frac{W_4}{L_4} = \frac{(W_6/L_6)}{2}\left(\frac{I_5}{I_7}\right) = \frac{(W_6/L_6)}{2}\left(\frac{(W_5/L_5)}{(W_7/L_7)}\right)$$

This gives the interrelation between the various device geometries. A starting point for the geometries can then be calculated from the equivalent circuit and the desired gain and frequency response. The g_m value of the devices is dependent on their geometries. The sizes of the buffer transistors will be of the same ratios to maintain the symmetry of the response. The actual sizes will be based on the frequency response and the current driving capability required.

The main drawback with this simple op-amp circuit lies in its frequency response. As there are three stages, each with an associated high-frequency cut-off value, the response is not in fact like the simple first-order response as shown in Figure 3.26. The additional poles to the response will be determined by the output resistance of each stage and the capacitance that it drives. For the first two stages, the output resistance will be high and so these will be dominant poles of the amplifier. It is found that with the high gain associated with the op-amp, the circuit is only conditionally stable and, if used in feedback circuits, oscillation can occur.

To make the circuit unconditionally stable, the amplifier can be internally compensated by connecting a compensation capacitor between the two high impedance nodes associated with the outputs of the first two stages. This is equivalent to putting the capacitor in a feedback configuration across the second stage, and has the effect of shifting the position of the dominant poles and making the overall circuit unconditionally stable. The value of the compensation capacitor is typically 5 pF.

3.7.5 Comparators

Comparators are usually two-input, single-output devices which compare the values of the two input voltages and produce an output close to either the positive rail voltage or the negative rail voltage, depending on whether input 1 was greater or less than the value of input 2. As the output voltage can easily be considered as a digital signal, comparators find most use in ADCs. Clearly an op-amp could be used in an open-loop configuration to realize a comparator circuit. The main requirements for a good comparator operation are high gain, a fast differential stage and a low offset voltage. As the device will be open loop, complexities of compensation in the op-amp are unnecessary and can slow the response, so comparators are often designed 'from scratch'. The circuit symbol, ideal voltage transfer characteristic and practical transfer characteristics, taking account of finite gain and offset voltage, are shown in Figure 3.29. The slope of the non-ideal transfer characteristic between v_{oL} and v_{oH} is directly related to the gain of the circuit, and the range of input voltage over which this transfer takes place represents the resolution of the comparator. This relates to the resolution, or number of bits, of an ADC formed using the comparator. For high resolution we need as high a gain as possible.

BJT circuits tend to have higher gain capability and also lower offset voltages than MOS circuits, so naturally make better quality comparator circuits. In general, the gain of the standard differential amplifier is insufficient and a second stage of amplification is required. As the comparator is usually only

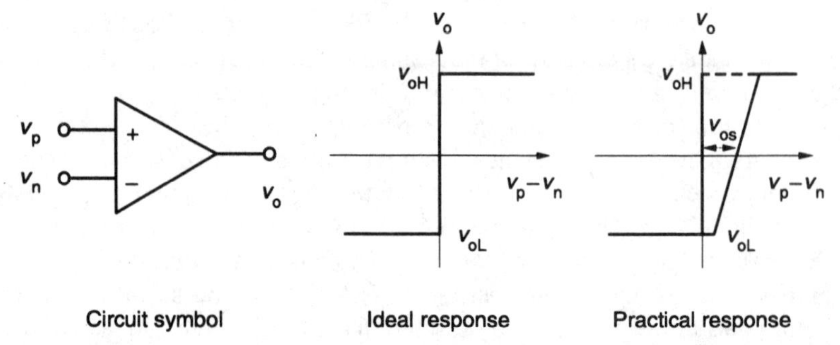

Figure 3.29 Circuit representation and characteristics of a comparator

lightly loaded, the output buffer circuit present in the op-amp is not required, so the circuit of a comparator is virtually identical to the first two stages of the uncompensated op-amp. The design criterion of having T_4 kept in saturation is the same in this case, so the design equations are similar to those outlined in the previous section. In this case however, the main design objective is high gain, and little consideration is paid to the frequency response of the device.

The offset voltage is the other major design constraint. This will arise from any asymmetry in the differential amplifier transistors. While the circuit can nominally be made completely symmetric, layout constraints and processing variations mean that the offset problem can never be completely eliminated at the fabrication stage. This problem can be eased by laying out the four transistors around a common centroid so that variations in layer thicknesses and doping densities, for example, cancel out as far as possible.

One way of increasing the gain of the comparator stage is to apply a system of positive feedback, or cross-coupling. This results in a comparator as shown in Figure 3.30. The differential stage in the centre now operates as a bistable or flip–flop circuit. Here we have to employ a digital clocking system as the circuit operates in two distinct modes at different times. Pass transistors are used to isolate different parts of the circuit during these phases. The capacitors C_1 and C_2 are monolithic devices which are used to store voltages between different clocking phases.

The first phase is the initialization or memory phase. During this phase clock ϕ_1 is high (V_{DD}) and clock ϕ_2 is low (ground). So transistor T_5 is switched off, no current can flow through the cross-coupled stage and so the bistable cannot operate. The pass transistors T_{10} and T_{11} are switched on, so that the two input signals v_p and v_n are stored on C_1 and C_2 respectively. The second phase is the decision phase. The clock ϕ_1 goes low, isolating the input signals, and clock ϕ_2 goes high, switching on transistor T_5 and activating the bistable. This part of the circuit will now switch, depending on the relative values of the voltages stored on the capacitors. Suppose, for example, v_p is greater than v_n, the voltage on the gate of T_2 is higher than that on the gate of T_1, and the bistable will regeneratively switch so that the voltage on C_1 rapid-

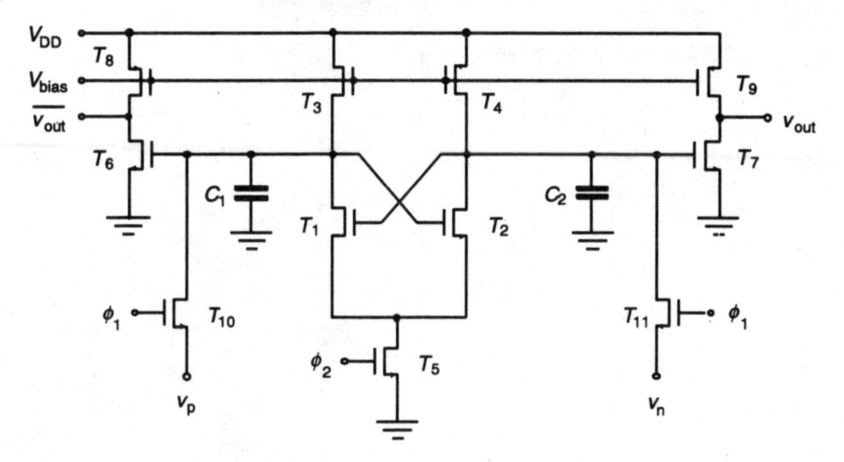

Figure 3.30 CMOS cross-coupled comparator circuit

ly goes to V_{DD} and that on C_2 goes to earth. So the output stages T_6, T_8 and T_7, T_9 ensure that v_{out} is high and $\overline{v_{out}}$ is low. The positive feedback action results in a much faster change in voltage than could be achieved purely from the action of the linear amplifier.

There still remains the problem of offset voltage which will ultimately limit the resolution of any ADC formed using the comparator. We have seen in the previous example that comparators can be operated in multiple phases, involving sampling, comparison and subsequent resetting. It is during the reset phase that we can incorporate techniques to account for the offset voltage. Such techniques go under the heading of *autozeroing*.

The basic autozeroing procedure is shown in Figure 3.31. In Figure 3.31a the comparator with the offset voltage is shown separately. During the reset phase (Figure 3.31b), the comparator is configured such that the offset voltage can be measured and stored on a capacitor. Then in the comparison stage, 3.31c, the capacitor is configured such that the offset voltage stored on it cancels the same offset voltage which is still present in the comparator.

A practical implementation, with the switching pass transistors and the two controlling clock phases, is shown in Figure 3.32. Here ϕ_1 and ϕ_2 are a non-overlapping, inverted pair of clock signals. During ϕ_1 the comparator is configured as in Figure 3.31b and the offset voltage is stored on *C*. During ϕ_2 the circuit is configured as in Figure 3.31c and the offset voltages cancel.

This relatively simple procedure will greatly ease the problems of comparator offset voltage and increase the resolution of the ADC, but it does not completely cancel all the offset. There still remain problems of charge injection from clock feedthrough in the switches, a high input resistance is required or charge will leak from *C* during ϕ_2, and there is also some dependence of the offset voltage on the common mode signal, so v_{os} may not be the same during the two phases. There is also the problem that the comparator is configured in

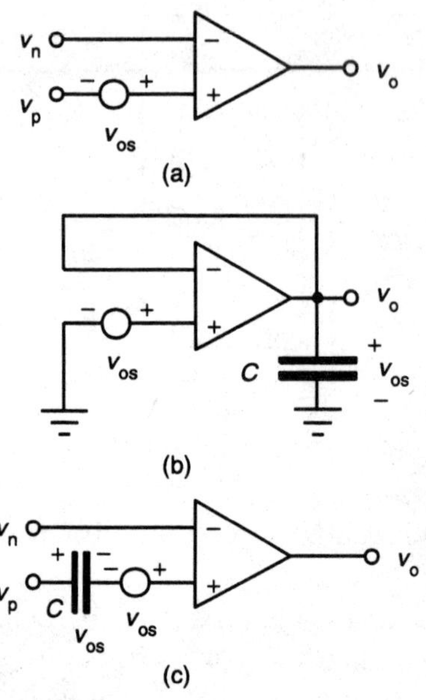

Figure 3.31 Autozeroing procedure for a comparator: (a) comparator with offset voltage; (b) circuit during reset phase; (c) circuit during comparison phase

a feedback mode, so may require compensation to guarantee stability. Other more complex autozeroing techniques and cross-coupling of comparators can improve the resolution, but always at the expense of larger circuits.

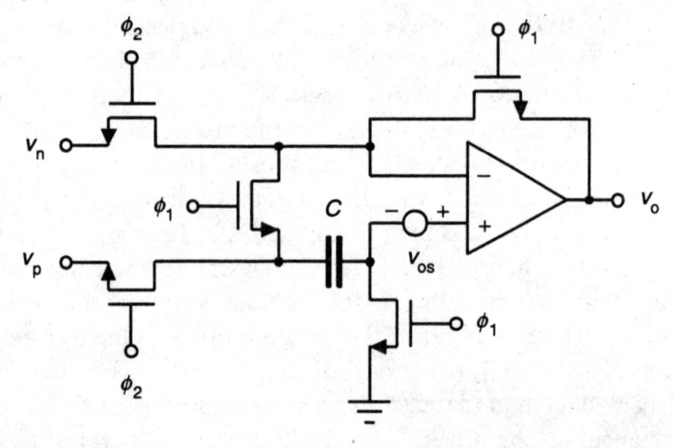

Figure 3.32 Implementation of the autozeroing procedure

3.7.6 Switched capacitor circuits

While most digital and analogue circuits were easily realizable with the development of monolithic integrated circuits and were rapidly transferred from the discrete form of circuit, often with great improvements in their performance, the one form of circuit that resisted this change was that of filters. Filters require resistor or inductor elements in combination with capacitors. While capacitors are easily realized monolithically, resistors are possible but at the expense of large circuit area and relatively poor stability. Inductors are virtually unrealizable except for very small values (of the order of nH) which are only useful at microwave frequencies.

Switched capacitor circuits were developed in the 1970s and 1980s to overcome these problems by the neat solution of using capacitors and high-frequency switching circuits to simulate the behaviour of resistors. The basic circuit of a switched capacitor is shown in Figure 3.33. With the switch initially in position A, the capacitor is charged to voltage V_1. If the switch is then

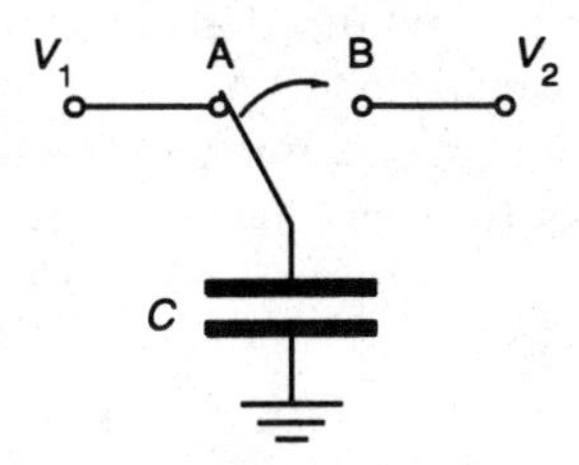

Figure 3.33 Basic switched capacitor circuit

moved to position B, the capacitor will now be charged to voltage V_2 and the amount of charge transferred during the operation between the two nodes is

$$Q = C(V_1 - V_2)$$

This process is repeated at a frequency f_s, so the amount of charge transferred per unit time is Qf and this is equivalent to a continuous current flow

$$I = Qf = Cf_s(V_1 - V_2)$$

So we have that the current is proportional to the potential difference, that is the switched capacitor is acting as a resistor with

$$R = \frac{V_1 - V_2}{I} = \frac{1}{Cf_s}$$

Provided that the switching rate is much higher than the signal frequencies in the circuit, the discrete switching process can be considered as a time contin-

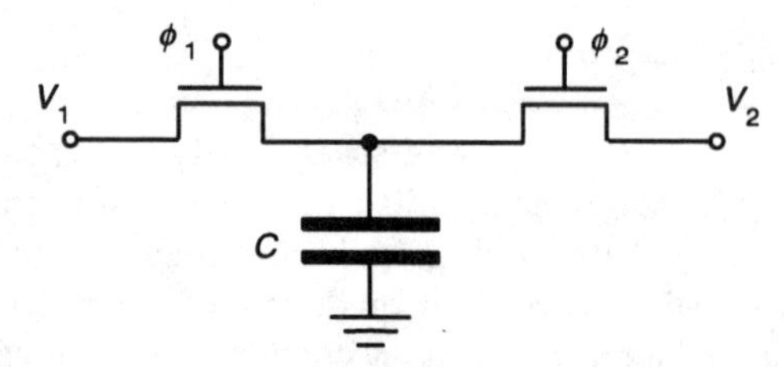

Figure 3.34 Practical implementation of switched capacitor resistance

uous process and the switched capacitor as a linear element. The practical realization of the switched capacitor (s-c) circuit using pass transistor switches and a two-phase non-overlapping clock is illustrated in Figure 3.34. Very high resistances are achievable, as it can be seen that the equivalent resistance is proportional to the inverse of the capacitance.

Once the basic resistor element has been realized, the whole range of associated circuits can be formed monolithically. Some brief examples will now be given. An integrator circuit based on an op-amp, and its s-c equivalent, are shown in Figure 3.35. The transfer function of the circuit is determined by the *RC* product, which in the s-c version is given by $C_f/C_s f_s$. This illustrates one of the important aspects of s-c monolithic circuits in that the transfer function is dependent on the ratio of two capacitances. It has already been pointed out that the ratio of capacitance values in ICs can be made more accurately than the absolute values of capacitance.

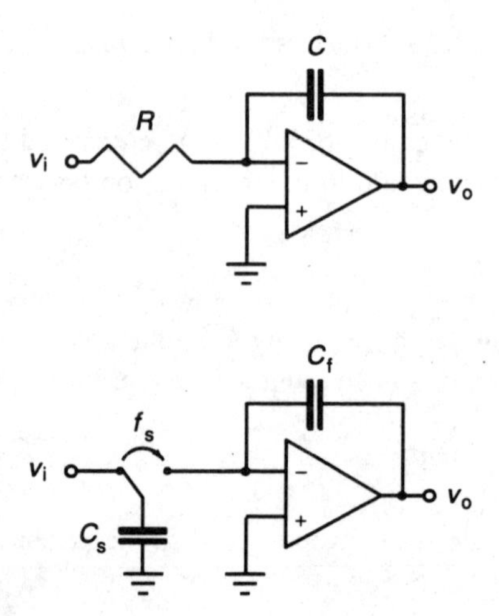

Figure 3.35 Simple integrator circuit and its switched-capacitor realization

We have already mentioned the difficulty of realizing inductors directly in monolithic form. However, with a s-c circuit the element now becomes realizable. If the voltages on either side of an inductor, L, are V_1 and V_2 then the current flowing through it is given by

$$I = \frac{V_1 - V_2}{\mathrm{j}\omega L}$$

We can use an integrator function to simulate this relationship, however in the integrator we only deal with voltages, so we must transform this equation by multiplying both sides by a normalising resistance R_n

$$V_n = IR_n = \frac{R_n}{\mathrm{j}\omega L}(V_1 - V_2)$$

The value of R_n is unimportant, so we can take it to be 1 Ω. In this case the equation above can be simulated by a using a differential input integrator with the capacitor ratio

$$\frac{C_f}{C_s} = f_s L$$

as illustrated in Figure 3.36.

We now have s-c circuits to simulate the necessary components and so filter circuits of high performance and stability can be generated on chip.

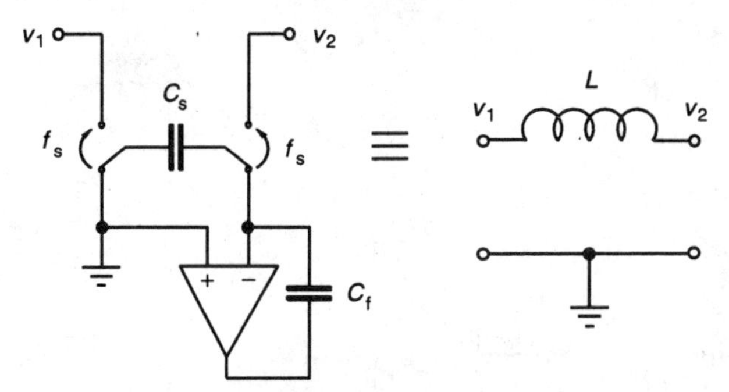

Figure 3.36 Switched-capacitor equivalent of an inductor

3.7.7 Analogue-to-digital and digital-to-analogue circuits

ADCs and DACs form the boundaries between the two modes of electronic data processing. However with s-c circuits and other analogue circuits making use of clock switching, and also with reliable CMOS and BiCMOS mixed signal ICs, there are now no clear demarcations between the analogue and digital world. ADCs and DACs are still very important building blocks and so should not be overlooked.

The requirement for these circuits is based on the fact that the digital processing of data and its mathematical manipulation can often be done more accurately, reliably and sometimes faster than by analogue processing. As most real-world signals, both input and output, are analogue in nature, there must be a way of representing the analogue signal in a binary form. The most common, and the one on which ADCs and DACs are based, is to have a time sequence of digital words, each word representing the sampled level (usually in voltage terms) of the analogue signal at that particular time. The number of bits in each word will define the total number of quantized levels at which the analogue signal can be represented. Usually the maximum analogue signal range is known, so the largest digital number (all 1s) represents this maximum, the smallest (all 0s) the minimum signal level, and the intervening range is divided linearly, so the number of levels will define the resolution to which the analogue signal can be represented. The more bits in a word, the finer the resolution.

Thus the general approach to ADC is to sample and hold the analogue signal at set time instances. The sampled level is then compared with a series of reference levels and, by using comparators, a digital sequence is derived that may be further encoded to get the correct word. The signal is sampled again and the next word generated. In the DAC process the digital word is taken, and according to the value of each bit, a summation of corresponding voltages is made to create the correct signal level.

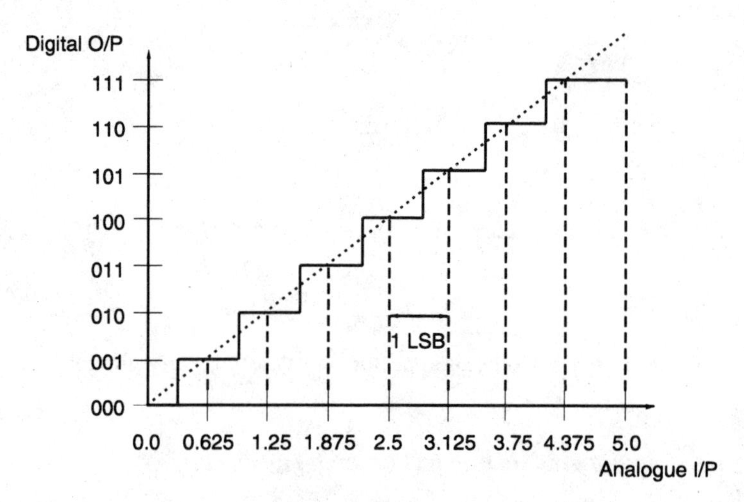

Figure 3.37 Ideal ADC characteristics

The ideal ADC process is represented by the transfer function in Figure 3.37, shown for a 3-bit (8-level) representation and a 5 V working range for the analogue signal. Each quantized level is centred on a fractional value of the full-scale signal value and has a width equivalent to the weighting of the least significant bit (LSB). The resolution of quantization accuracy is therefore described as being $\pm \frac{1}{2}$ LSB. There are of course many errors in real systems that will cause a deviation from this ideal case, and some of the more important of these are illustrated in Figure 3.38.

Figure 3.38a illustrates offset error where there is an offset voltage present in the conversion process and the whole characteristic is shifted along the x-axis. Here the resolution of each level remains at the correct $\pm \frac{1}{2}$ LSB, but the levels are not centred on the correct input values. Figure 3.38b illustrates gain error where the slope of the transfer curve does not give the correct mapping of values. Here there will be a progressive error both in the position and width of the quantization levels. Figure 3.38c illustrates a non-linearity termed integral non-linearity, where the transfer function is now no longer a straight line,

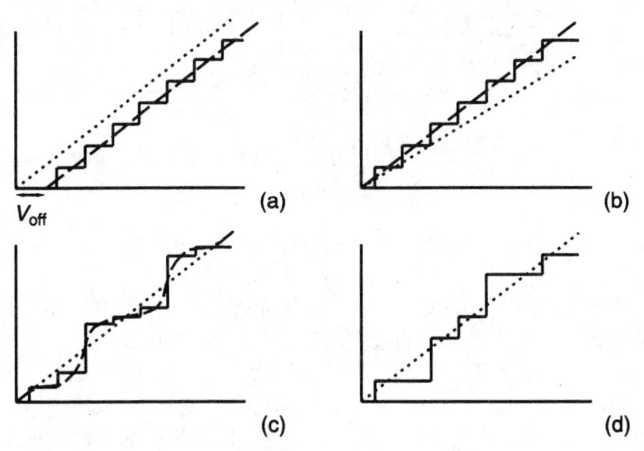

Figure 3.38 Errors in ADC characteristics: (a) offset error; (b) gain error; (c) integral non-linearity; (d) differential non-linearity

again causing variations in position and size of the quantization levels. Figure 3.38d illustrates another non-linearity, termed differential non-linearity, where the transitions in the levels are misplaced. This error can sometimes be so large that complete transitions are missed out, as shown. These various errors can occur simultaneously at different degrees, and it is often difficult to resolve the source of the non-idealities in a practical ADC.

There are three basic approaches to ADCs – serial, successive-approximation and parallel (flash). We shall look briefly at each of these systems in turn.

Serial ADC

This circuit is also known as an integrating ADC, as it is based on an integrator that has a reference voltage as its input. The output of the integrator is compared with the analogue input voltage, and the output of the comparator is used to control a digital counter circuit. The system is illustrated in Figure 3.39. The integrator circuit has an output that is a ramped voltage, and the start of the integration cycle coincides with the start of the digital counter. Once the voltage V_i exceeds the analogue voltage V_{in}, the comparator outputs an edge to stop the counter. The count number that the counter had reached by this time is a representation of the analogue signal level. The larger the value of V_{in}, the longer it will take for V_i to exceed it, the longer the counter will operate and hence the higher will be the resulting digital word.

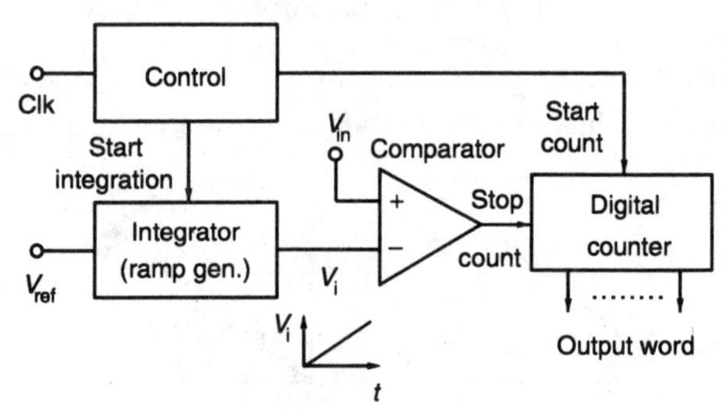

Figure 3.39 Serial ADC

A variation of this approach is to use a dual slope integrator and perform the counting operation in two stages. Firstly, V_{in} is connected to the integrator input and the integrator operated in the positive direction for a known period. V_{ref} is then connected to the integrator input and the integrator operated in the negative direction, and the point where $V_i = 0$ is detected by the comparator and the counter stopped. The advantage of this approach is that it eliminates the dependence of the time count on the gain of the integrator, and potentially leads to very accurate conversion.

Successive-approximation ADC

The problem with the serial ADC is that the endpoint of the conversion is dependent on the value of V_{in} and so is not known in advance of the conversion. Integrating this with a larger system may require the output word to be held in a register until a point in time corresponding to the conversion of the maximum value of V_{in}. The successive-approximation ADC avoids this problem as it performs the conversion in a fixed period, depending on the number of bits in the converted word. The outline of the circuit is shown in Figure 3.40. The DAC provides a reference voltage to be compared against V_{in} which is some fraction of V_{max}. At the first clock cycle this reference voltage is set at $V_{max}/2$. This will generate the most significant bit (MSB) of the converted word. If V_{in} is greater than $V_{max}/2$, the MSB is 1. This is stored for output at

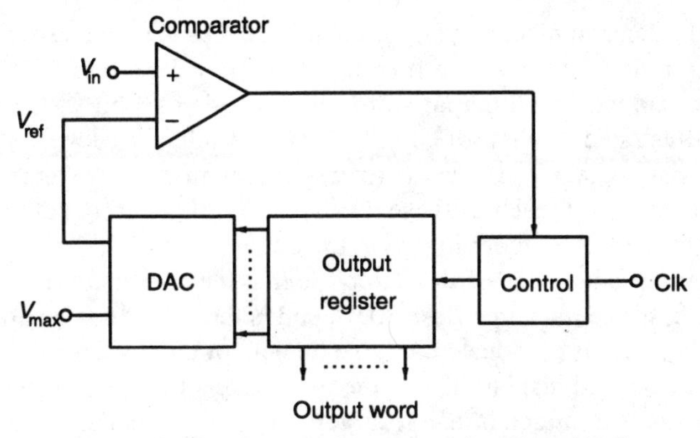

Figure 3.40 Successive-approximation ADC

the end of the conversion, but is also used to determine the next value of the reference voltage. If the MSB is 1 then the next value is $\frac{3}{4}V_{max}$, and if it is 0 then the next value is $\frac{1}{4}V_{max}$. This value is applied to the comparator and the comparison made during the next clock cycle. The process is repeated for the number of bits in the converted word, and at each stage the reference voltage 'homes in' closer to the value of V_{in} as illustrated in Figure 3.41 (here $V_{in} = 0.6V_{max}$, the 3-bit output is 100). So the successive-approximation ADC operates on a fixed conversion time which is usually much shorter than that of the serial ADC.

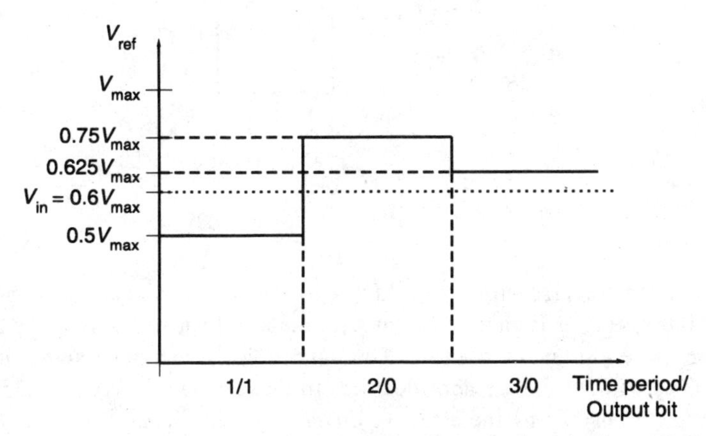

Figure 3.41 Operation of successive-approximation ADC

Parallel or flash ADC

In the parallel converter, the input voltage is compared with the reference voltage corresponding to each quantization level simultaneously. This therefore requires the same number of voltage comparators as there are levels, each one producing a digital bit that can then be passed to a 2^n to n encoder circuit to derive the correct digital output word. An example of 3-bit (8-level) parallel ADC is illustrated in Figure 3.42. The voltage reference levels are derived using a resistor chain to divide the reference potential into the correct values. V_{in} is compared with each of these levels, the eight resulting bits being input to the eight to three encoder to derive the output word.

The obvious advantage of this circuit is that the conversion is done in one clock cycle, hence the name flash ADC, and is the fastest conversion method. The disadvantage is obviously the large amount of circuitry required, increasing with the desired resolution, and the requirement to match the behaviour of the components to maintain the accuracy. The process is therefore only used where speed is the overriding consideration.

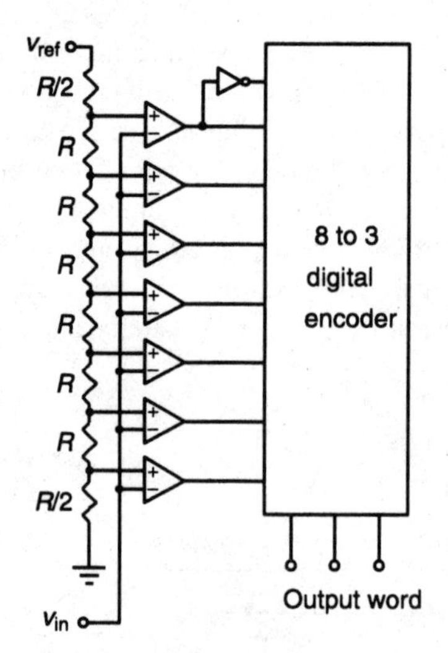

Figure 3.42 Parallel (flash) ADC

DACs are the reverse process of ADCs. The ideal transfer characteristic is very similar to that of Figure 3.37, but with the x and y axes interchanged, the input being the digital word and the output being the analogue signal level. The DAC can suffer similar non-idealities to those shown in Figure 3.38, the only real difference being the effect of differential non-linearity which in this case leads to a non-monotonic characteristic, whereby an increase in the value of the binary word does not always lead to an increase in value of the analogue signal level.

DACs also come in serial and parallel forms, each of which will now be briefly described.

Serial DAC

The serial DAC is based on a summing process, taking the effective bit weight or equivalent value of each bit in turn and adding that contribution if the bit is a 1, but not if the bit is a 0. The basic circuit for this DAC is shown in Figure 3.43. The two capacitors have the same value and the process involves charge sharing between the two to get the successive divide-by-two weighting for each bit. The first step of the operation is to discharge C_2 by closing switch S_3 while S_2 is open. S_3 is then kept open through the conversion process. The bits are considered, starting with the LSB and progressing through to the MSB. If a particular bit is a 1 then switch b is closed and $\bar{b}$ is open. If the bit is a 0 then b is open and $\bar{b}$ is closed. S_1 and S_2 are operated on a two-phase non-overlapping clock sequence.

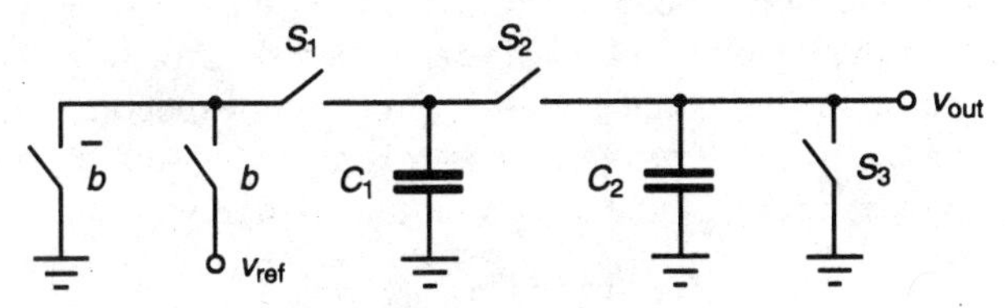

Figure 3.43 Serial DAC circuit

The sequence at each step is:

(1) The b switches open and close;
(2) S_2 is opened, isolating C_1;
(3) S_1 is closed, charging C_1 to V_{ref} or 0 V, depending on the bit value;
(4) S_1 is opened and S_2 is closed – the charge stored on C_1 is shared between the two capacitors, and the weight of the bit stored on C_1 is divided by two;
(5) S_1 is opened and the process repeats.

At each step the weight of each of the stored bits is divided by two, so at the final step when the MSB is distributed, each bit will have the correct value and the total charge, and therefore the voltage on C_2 will be the analogue representation of the bit train and the conversion is complete.

The conversion will take the same number of time steps as the number of bits in the word (plus a reset and readout step), since the bits are dealt with in a serial manner. The circuit is very simple however.

Parallel redistribution DAC

The parallel system for a DAC is through the use of a scaling network which is under control of the input digital word. This steers the correct weight of some parameter to a summing point which provides the correct analogue signal level. The parameter involved can be current, voltage or charge, or even a combination of these, each with its own particular advantages. We shall just describe a simple charge distribution system based on weighted capacitor values and converting a 3-bit input number. The system described can be extended to longer binary inputs.

The circuit is shown in Figure 3.44 and, like the serial DAC, is based on a switch system operated by a two-phase non-overlapping clock system and switches controlled by the input bit values. There are only two phases to the system, so the conversion is done within one cycle. In the first phase (ϕ_1), V_{out} is reset to zero as the top plates of all the capacitors are grounded. The bottom plates of the capacitors are connected to V_{ref} if the corresponding bit is a 1, or grounded if the bit is a 0. Therefore charge is stored on the capacitor array according to the pattern of 1s and 0s and on the relative weight of each bit.

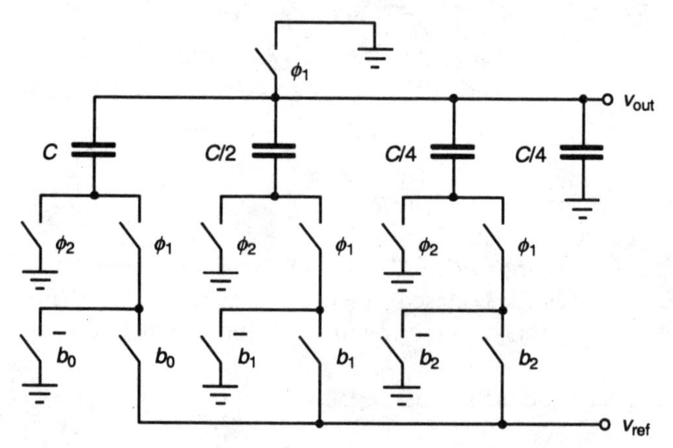

Figure 3.44 Parallel charge redistribution DAC

In the second phase, the bottom plates are grounded and the top plates connected together. There is no loss of stored charge, as this charge redistributes itself among the capacitor array which always has a total value of $2C$. Hence the charge that was proportional to the weighted value of the digital word is converted to a voltage, V_{out}, which is also proportional to this weighted value, and hence the analogue equivalent of the digital word.

Being a parallel process, the conversion time is very quick, but again at the cost of circuit complexity. This grows as the number of bits is increased, with the added complications of maintaining accurate capacitor ratios and the exponential growth in the value of the largest capacitor.

References

Geiger, R.L., Allen, P.E. and Strader, N.R. (1990) *VLSI Design Techniques for Analog and Digital Circuits*, McGraw-Hill, New York.

Widlar, R.J. (1971) New developments in IC voltage regulators, *IEEE Journal of Solid-State Circuits*, Vol. SC-9, No. 1, pp. 2–7.

Bibliography

L.E.M. Brackenbury, *Design of VLSI Systems – A Practical Introduction*, Macmillan, Basingstoke, 1987.

T.E. Dillinger, *VLSI Engineering*, Prentice-Hall, Englewood Cliffs, New Jersey, 1988.

E.D. Fabricius, *Introduction to VLSI Design*, McGraw-Hill, New York, 1990.

P. Gray and R. Meyer, *Analysis and Design of Analog Integrated Circuits*, Wiley, New York, 1993.

R. Gregorian and G.C. Temes, *Analog MOS Integrated Circuits for Signal Processing*, Wiley, New York, 1986.

P.J. Hicks, *Semi-Custom IC Design and VLSI*, Peter Peregrinus, London, 1983.

J. Mavor, M.A. Jack and P.B. Denyer, *Introduction to MOS LSI Design*, Addison-Wesley, London, 1983.

C. Mead and L. Conway, *Introduction to VLSI Systems*, Addison-Wesley, Reading, Massachusetts, 1980.

M.J. Morant, *Integrated Circuit Design and Technology*, Chapman & Hall, London, 1990.

A. Mukherjee, *Introduction to nMOS and CMOS VLSI Systems Design*, Prentice-Hall International, Englewood Cliffs, New Jersey, 1986.

D.A. Pucknell and K. Eshraghian, *Basic VLSI Design*, 3rd edn, Prentice-Hall, Sydney, Australia, 1994.

C.J. Savant, M.S. Roden and G.L. Carpenter, *Electronic Design Circuits and Systems*, 2nd edn, Benjamin/Cummings, Redwood City, California, 1991.

A.S. Sedra and K.C. Smith, *Microelectronic Circuits*, 2nd edn, Holt, Rinehart and Winston, New York, 1987.

J.P. Uyemura, *Fundamentals of MOS Digital Integrated Circuits*, Addison-Wesley, Reading, Massachusetts, 1988.

H.J.M. Veendrick, *MOS ICs – From Basics to ASICs*, VCH, Weiheim, Germany, 1992.

Questions

3.1. Consider Figure 3.2 with an input voltage $V_i = 0.2$ V (logic 0). Calculate the output current flowing when the output is shorted to ground. Assume a high value of β. What is the minimum value of β for which this analysis will hold? Comment on the function of the 130 Ω resistor. [33.75 mA, 14]

3.2. For the ECL circuit of Figure 3.6, with transfer characteristics of Figure 3.7, calculate the upper and lower noise margins if the switching definitions are:
(a) 90 per cent of I_E;
(b) 99.9 per cent of I_E. [(a) 0.475 V, 0.355 V; (b) 0.357 V, 0.237 V]

3.3. In an analysis of an nMOS technology with depletion mode loads, it is found that the optimum W/L sizes are 12 μm/6 μm for the driver transistor and 6 μm/12 μm for the load transistor. Calculate the transistor sizes for 3-input NOR and 3-input NAND gate configurations.

[NOR: 12 μm/6 μm, 6 μm/12 μm; NAND: 12 μm/2 μm, 6 μm/12 μm]

3.4. Four identical nMOS pass transistors, with $V_T = 1$ V, are cascaded source to gate, with the gate of the first transistor connected to V_{DD} (5 V). What is the highest available output voltage from the output of the fourth transistor? Comment on the result. [1 V]

3.5. A signal that normally drives a 2 pF input gate capacitance is to be taken off chip, driving a 500 pF load capacitance. Calculate the number of buffer stages required for minimum overall delay. If only four stages are to be used, calculate the capacitance ratio between stages. Estimate the increase in delay for this non-ideal case. [6, 3.98, 6 per cent]

3.6. For a 1 μm CMOS technology, an inverter is designed with $K = 2$. Calculate the transistor sizes if:

(a) the n-channel device is of minimum size;

(b) the p-channel device is of minimum size.

[(a) $W_n = 1$ μm, $L_n = 1$ μm, $W_p = 2$ μm, $L_p = 1$ μm; (b) $W_n = 1$ μm, $L_n = 2$ μm, $W_p = 1$ μm, $L_p = 1$ μm]

3.7. Using the equations in section 3.7.3, calculate the resistor ratio R_2/R_3 required to provide a reference voltage with zero temperature drift of 1.5 V, given that the current ratio I_1/I_2 is 10 and the value of V_{BEO} is 0.65 V at 290K. [14.8]

3.8. Derive the output from a successive-approximation ADC if the input is 0.3 V_{max} and the output is:

(a) in 3 bits;

(b) in 8 bits. [(a) 010; (b) 01001100]

4 IC Realization
How does it come together?

4.1 Introduction

So far we have looked in detail at types of IC families that are available and also at the building blocks that we can use for both analogue and digital circuit design. In this chapter we shall look at the various ways that the integrated circuit as a whole comes together. At this level of circuit description, there are two classifications of IC. Firstly, there are universal ICs that can be used for any number of different applications. These devices range from the individual gate packages, such as the TTL 7400 series, through to the various microprocessor chips and also memory ICs. On the other hand, there are ICs that are designed to perform a specific function, termed application specific ICs (ASICs).

Historically, ASICs are a relatively new development, since originally the application of a circuit was introduced at the board level, using a particular combination of the universal ICs as building blocks. Nowadays the combination is largely made at the chip level and ASIC use is growing rapidly. The universal ICs still find their uses: the gate packages are still required for small amounts of 'glue logic' in large systems, for prototyping and for small circuit realization. The larger universal ICs, such as microprocessors and memory devices, cannot at present be integrated into ASICs in a practical way, so the integration of these ICs to the system remains largely at the board level. This situation is changing, however, and such circuits will soon be available as part of an ASIC.

The descriptions so far have been concerned with digital ICs. A similar situation exists for analogue circuits; here the universal ICs are primarily building blocks such as op-amps, comparators, oscillators, etc. There are no real analogue equivalents of the large-scale ICs such as microprocessors and memory devices. Analogue ASICs are available, but as described later in the chapter, the options in this area are more limited than in the digital area. As far as mixed signal ICs are concerned, these devices are entirely ASIC in form.

This chapter will first briefly describe some of the universal ICs, but the main emphasis will be on ASICs, as these are more likely to be the type of circuit that modern IC designers will work with, the design of microprocessors and memory elements being somewhat specialized work. The particular ASIC approach that a designer takes will depend on the available technology, the circuit application, the number of circuits of a particular design to be generated and the financial considerations. These various aspects were covered in Chapter 1, so this chapter will only discuss the available technologies and design approaches.

4.2 Universal ICs

As described in section 4.1, universal digital ICs exist at the two ranges of complexity – either basic gate packages or large-scale microprocessors and memory arrays.

4.2.1 Gate packages

The operation of gate circuits and the different technologies available have been covered in Chapters 2 and 3, so the material will not be repeated here. The two most common technologies that are implemented as gate packages are TTL and CMOS, both available in the familiar 74 series. Also becoming prominent are the ECL gates in the 10K and 100K series.

Individual gates often only require a small number of inputs and a single output, along with power and ground lines, and in the case of sequential gates a clock input and often reset and clear lines. Therefore the number of pins on these packages need only be very small. Often several gates are put into the same package, sharing power lines. For example, the 7400 IC consists of four two-input NAND gates in a 14-pin dual-in-line (DIL) package, as shown in Figure 4.1. A sample list of some of the packages in this family is given in Table 4.1.

Table 4.1 *Partial list of 74 series of digital gate packages*

Number	*Function*	*Number*	*Function*
7400	Quad 2-input NAND	7430	8-input NAND
7402	Quad 2-input NOR	7432	Quad 2-input OR
7404	Hex inverter	7442	BCD-decimal encoder
7408	Quad 2-input AND	7470	J–K flip–flop
7410	Triple 3-input NAND	7474	D-type flip–flop
7411	Triple 3-input AND	7483	4-bit full adder
7420	Dual 4-input NAND	7490	Decade counter
7421	Dual 4-input AND	7491	8-bit shift register
7427	Triple 3-input NOR		

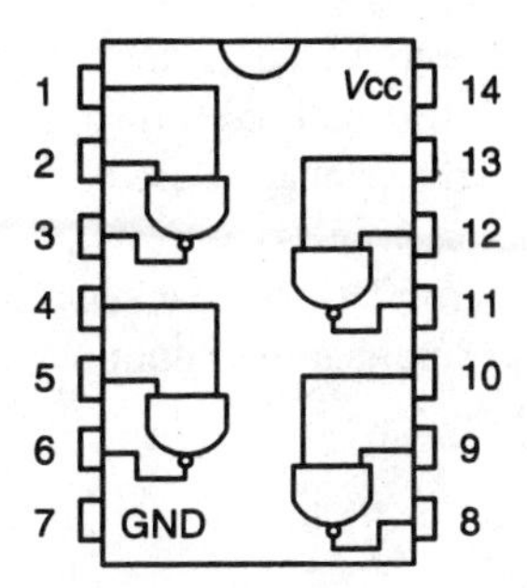

Figure 4.1 7400 quad two-input NAND package

4.2.2 Microprocessors

The heart of the digital computer is the processor unit. Very basically this consists of a central arithmetic logic unit (ALU) which performs the calculating functions, registers to store the digital words being operated on, and some memory area to store the instructions that the processor must obey. In addition there are interfaces to the outside world, and clocking and control circuitry. The architecture of a typical microprocessor is shown in Figure 4.2. The detailed description of the design and operation of these circuits is beyond the scope of this text, but is well covered in many other modern texts (see Bibliography at end of chapter).

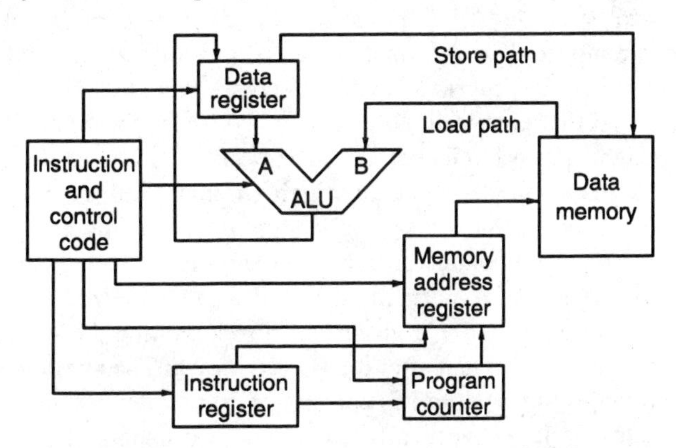

Figure 4.2 Typical microprocessor architecture

Improvements in microprocessor performance have centred around the architecture of the ALU so that arithmetic operations can be performed in the minimum number of clock cycles. In addition, the width of the data paths has grown steadily from 8 through 16 and 32 to the state-of-the-art figure of 64, although 32 is the standard commercially available size. A larger number of bits being operated on by the processor at any one time can result in a more accurate, or higher-resolution, floating-point operation. However, the large width data buses mean that the packages used to mount the ICs must have a very large number of pins, often in excess of 100. The traditional DIL package is not practical, and square, flat-pack packages are used with smaller pitch pins which must first be mounted into chip carriers that are themselves attached to the circuit board.

4.2.3 Memory circuits

Memory integrated circuits are large regular arrays for the temporary or permanent storage of digital information, each bit being stored in a single cell. The improvements in computing performance, coupled with the growing demands for electronic data processing and storage, have meant that the size of these memory circuits has grown almost exponentially, and each can now store many millions of bits. Therefore the pressure on the memory circuit designer has been to minimize the area of each cell, so that the overall chip size does not become impractical. This minimization has taken place both through the improvements in processing technology, in reducing the circuit feature sizes, as well as through novel cell designs.

Memory circuits fall into three broad areas. In read only memory (ROM), the digital bits are stored on a long-term basis, and rarely, if ever, altered. Such memories are used, for example, in the instruction set for a microprocessor. ROM tends to be non-volatile, that is the stored information is not lost when the power is removed. The information can be put into the memory in a number of ways. It may be programmed in at the mask level by tailoring the metallization layer so that each cell stores a 1 or a 0. Once the ROM IC has been manufactured the data is totally permanent and cannot be altered. Programmable ROM (PROM) is available, one type utilizing miniature fuses associated with each cell which can be 'blown' by application of a high voltage.

In reality, these semiconductor fuses perform the reverse action of normal wire fuses. A wire fuse is normally a low-resistance element until the high voltage is applied, and then a large current flows, melting the wire and resulting in an open-circuit element after it has been blown. The miniature fuse used for programming semi-custom ICs consists of an extremely thin, small cross-sectional dielectric layer, vertically separating two conducting layers (metal and/or polysilicon) – it is therefore a normal high-resistance element. On the application of a sufficiently high-voltage pulse, the dielectric layer will break down and a relatively high, short-duration current will flow. This is sufficient to melt part of the conductor and generate a low-resistance link between the two conducting layers. Because of the reverse action of this semiconductor fuse, it is sometimes termed an antifuse. The fuse blowing action, like its wire counterpart, is however irreversible.

The designer blows or does not blow a particular cell according to whether a 1 or 0 needs to be stored at that location. Once blown, the ROM is permanently programmed. By using floating gate structures in MOS technology, which can store charge on a long-term basis, erasable PROM (EPROM) circuits are realizable. When first introduced, EPROMs had to be erased by exposure to ultra-violet light which alters the resistance of the dielectric material surrounding the gate, lowering the resistance and allowing the charge to dissipate; the ROM can then be reprogrammed. Nowadays the erasing process can be done electronically, so we now have electronically erasable PROM (EEPROM) circuits. However, as the linking 'fuse' in this case is a MOS transistor structure, the area is very much greater than the miniature antifuse, so erasable circuits tend to have a larger circuit area than the singly programmable equivalent.

In random access memory (RAM), the system reads and writes to it on a regular basis, for example for storing the computer program that a system is currently running, or for the storage of data on which the program is operating. RAM tends to be volatile in that the stored information is lost when the power supply is disconnected. It can be further separated into two forms. In dynamic RAM (DRAM), the charge, which represents the stored bit, will naturally leak away and the information will be rapidly lost; the cell has to be constantly refreshed, usually on every clock cycle. DRAMs can be as simple as a single FET device per cell, the charge being stored on the gate capacitance, and so they tend to have the highest packaging density of all memory ICs. Static RAM (SRAM) can retain the bit in the cell without the need for refreshment, provided of course that the main power supply is maintained. These are usually based around a bistable circuit, and hence are more complex and have a larger area than DRAMs.

The third classification is that of serial memories, which tend to be more specific in their use, for example in shift registers or charge coupled devices (CCDs).

4.2.4 Analogue circuits

Universal analogue ICs only exist at the equivalent of the digital gate level. Some of these circuits have already been described, from a transistor level design point of view, in Chapter 3. The typical circuits commercially available are operational amplifiers, comparators, oscillators, ADCs and DACs, analogue switches and display drivers. Like the digital circuits, they are available in both bipolar and MOS technologies.

4.3 Programmable logic devices

As indicated in section 4.1 , the main interest for the modern IC designer is in the area of ASIC design. There are a number of ASIC approaches, which will be detailed in this and subsequent sections, and which have different levels of complexity and therefore flexibility. The simplest form is that of the programmable logic array (PLA).

4.3.1 Basic PLA layout

The basic form of the PLA is illustrated in Figure 4.3. It consists of two main areas termed the AND plane and the OR plane. These terms derive from the

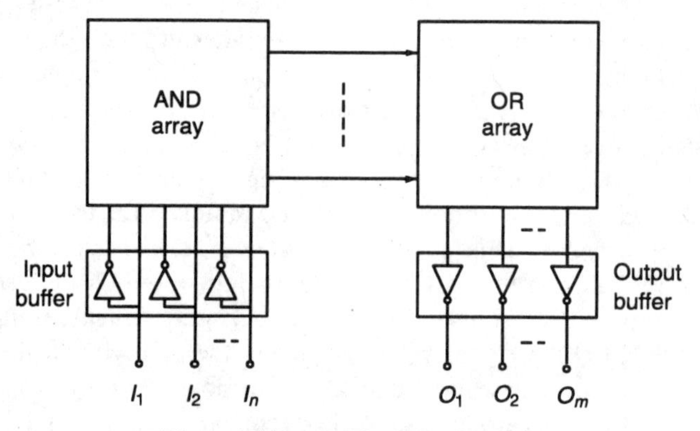

Figure 4.3 Basic PLA layout

fact that the fundamental form of the PLA is to generate any Boolean function that is represented in its sum-of-products form. The AND plane forms the product terms (often the prime implicants) and these are then summed in the OR plane. This operation can best be illustrated by means of an example. The Boolean function we shall derive is

$$f = \overline{A}\,\overline{B}C + \overline{A}B\overline{C}$$

Figure 4.4 illustrates the most common nMOS implementation of the PLA. The AND and OR planes consist of grids of conductors with FETs located at particular crossing nodes. The gate and drain are connected to the crossing conductors and the source is connected to ground. Note that the AND and OR planes are identical in structure, but rotated 90° to each other. Consider now the top row in the AND plane. When a logic 1 is applied to the gate of any of the FETs, that particular device will be turned on, the drain line will be pulled down towards earth and the logic value on the row will be a 0. Thus the

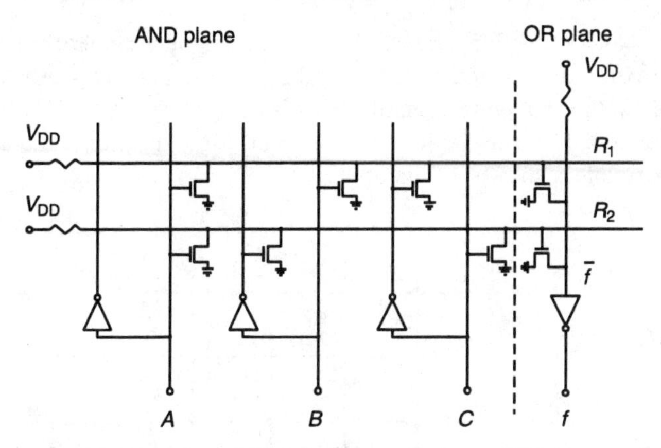

Figure 4.4 nMOS implementation of PLA

Boolean function represented by the top row is the NOR function of those inputs to which transistor gates are attached. Hence

$$R_1 = \overline{(A + B + \overline{C})}$$

and by DeMorgan's law this is

$$R_1 = \overline{A}\,\overline{B}C$$

So although the structure strictly generates a NOR function, this is equivalent to an AND function on the inverse of the inputs. Likewise, examining the position of the FETs in the second row, we have

$$R_2 = \overline{(A + \overline{B} + C)} = \overline{A}\,B\,\overline{C}$$

Now if we look at the OR plane, as noted this is the same structure, so again will generate the NOR function of those variables that have a FET present on the line. In this case both row functions have FETs and so will be combined as $\overline{(R_1 + R_2)}$. This is actually the inverse of the final function that we require, so the output buffer stage is an inverter and the true function *f* is generated. Strictly the OR plane is made up of the NOR function of the array in combination with the inverting output buffer.

It can be seen that any Boolean functions can be generated in this fashion. The number of input lines depends on the total number of input variables in the function(s). The inverses of these inputs are also required as data lines within the AND array. This array is programmed by placing FETs at nodes according to the *inverse* of the input variables. The number of rows required depends on the total number of product terms; these rows then form the input rows to the OR plane. The number of columns in this array is determined by the total number of separate Boolean functions, or outputs, to be generated. The programming of the OR plane involves placing FETs at the nodes corresponding to the product terms required in the output function.

If the Boolean functions are available as a truth table rather than a sum-of-products expression, then the PLA can still be programmed directly, as will be illustrated in the following example. This is the realization of a two-input half adder function. The truth table of the sum-and-carry functions, and the resulting PLA circuit is shown in Figure 4.5.

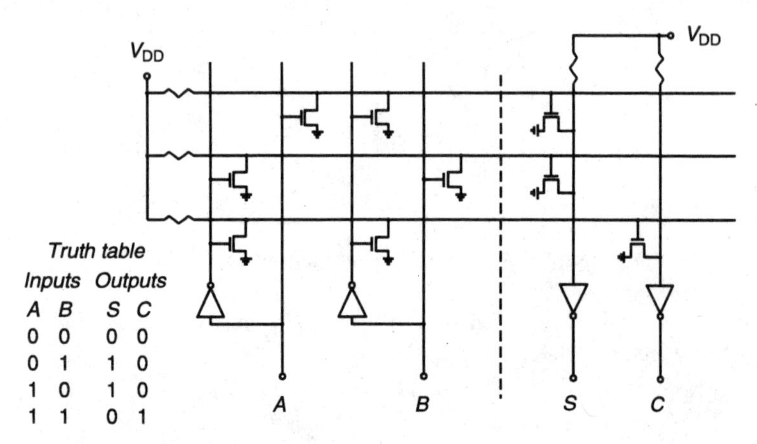

Figure 4.5 PLA implementation of half adder function

The first point to note is that the number of rows is equivalent to the number of lines in the truth table, with the exception that a row does not need to be included if *all* the outputs in the corresponding truth table line are 0. This is the case for the first row of the truth table in this example, so there is no corresponding row in the PLA. If the row is absent for that input combination, the OR plane columns are pulled up to V_{DD} and the outputs are 0 by default. The programming of the PLA is a direct mapping of the pattern of 1s and 0s in the truth table to give the location of the FETs. For the AND plane, the position of the FETs corresponds to the complementary position of the input variables, that is if there is a 1 in the truth table, the FET is placed on the inverse line. If there is a 0 in the truth table, the FET goes on to the true function line. So the pattern of 0, 1, 1 reading down the *A* column of the truth table maps to FETs on A, $\overline{A}$, $\overline{A}$ respectively. For the OR plane, the presence of a FET corresponds to a 1 in the output column of the truth table. If there is a 0 in the output, the FET is omitted.

By use of these simple rules, it is easy to programme a PLA from either the Boolean functions or the truth tables. In general, the minimized Boolean form will provide the smallest array sizes, so the choice of starting point for the programming may depend on the size of the PLA that the designer has available. The programming is usually automated and computer controlled (see Chapter 5), so the designer need only worry about the size aspects.

4.3.2 Variations of the PLA

The architecture of the basic PLA can be adapted to a PROM architecture. In this case the AND array acts a decoding block, the n inputs being the digital word corresponding to the row in the OR array in which the stored word is retained. The word is m bits wide, m being the number of output lines. There are 2^n rows in the arrays, so the AND plane is a full decoder and has no requirement for programmability; only the OR plane is programmed, with the required data being stored in this section.

An alternative is to have only the AND plane programmable, the OR plane being fixed, so only a certain combination of row functions is available. This variation is termed programmable array logic (PAL). The three variants are summarized in Figure 4.6.

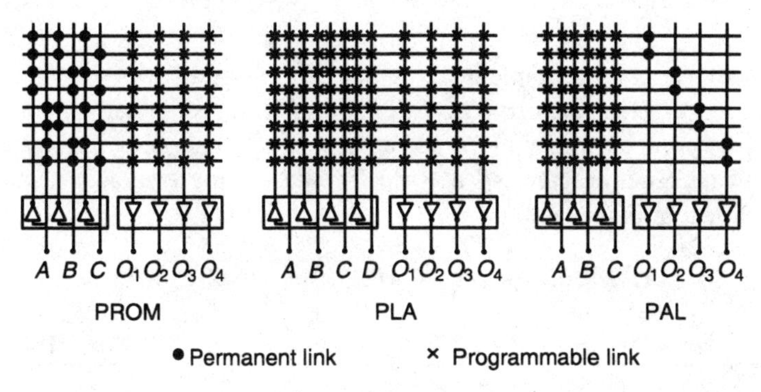

Figure 4.6 Comparison of PROM, PLA and PAL architectures

4.3.3 Sequential logic in PLAs

The circuits described above have all been concerned with the implementation of combinational logic via Boolean algebra or truth tables. It is possible, however, to incorporate sequential logic into the PLA architecture in order to realize such circuits as finite state machines (FSMs) and shift registers. To generate a simple FSM, the standard PLA layout can be adapted as illustrated in Figure 4.7. To achieve this, the sequential control must be made using a two-phase clocking system in order to control the timing of the feedback signals.

Other generalized sequential circuits can be realized using this type of architecture by incorporating flip–flops into the output buffer from the OR plane and to have the ability to feed back the outputs from these gates into the input lines of the AND plane. In this case the clock control is made via the clock inputs to the flip–flops, so only a single-phase clocking system is necessary. Often there is sufficient flexibility in the AND plane programming of

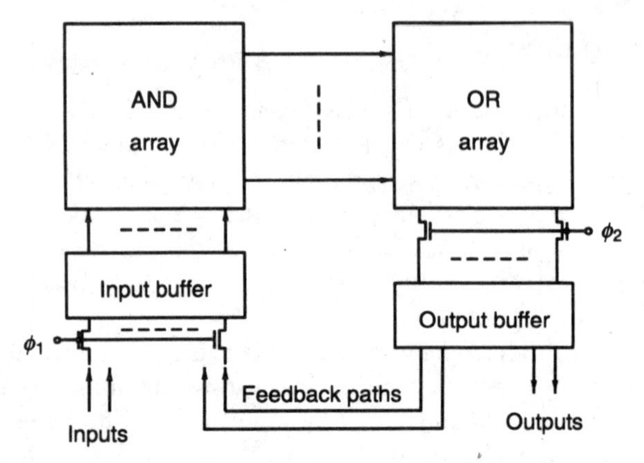

Figure 4.7 Finite-state machine using a clocked PLA

the fed-back variables to implement the device as a variation of the PAL, with a permanent OR plane pattern. This enables simplified computer generation of the required programming pattern and control of the 'chip-blowing' hardware to take place. The structure of a typical commercially available sequential PAL is shown in Figure 4.8.

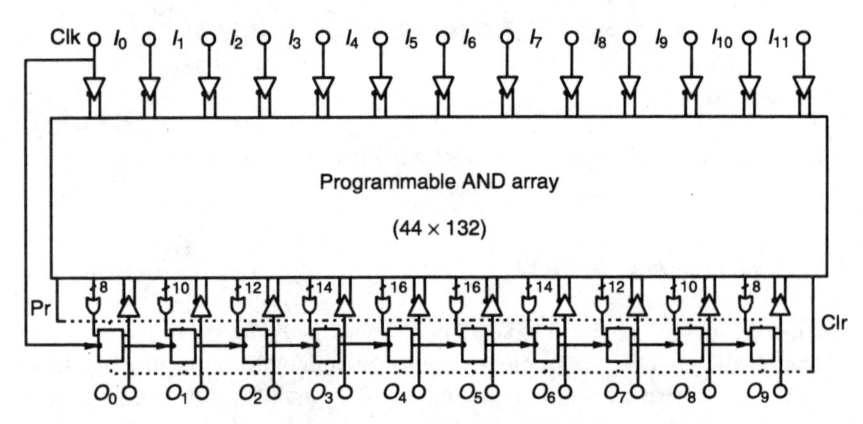

Figure 4.8 Block diagram of typical sequential PAL

4.4 Gate arrays

The next level in the 'semi-custom' approach to IC design is the gate array. As the name suggests, these circuits consist of an array of pre-formed gates, the designer having control over how the gates are connected. There are two main approaches to this – mask programmable and field programmable gate arrays (FPGA). In the former, the customization is done in the design of one or more of the metallization layers of the mask set. Once the circuit has been fabricated, the pattern is set and irreversible – this is the same situation as the basic mask programmed PLA and ROM. The field programmable approach uses the blowable 'fuse' approach, as in the EEPROM and erasable PLA.

4.4.1 Mask programmable gate arrays

The mask programming approach to the gate array provides the designer with more flexibility at the cost of the design being a 'one-off'. In this case the circuit is complete in its design with the exception of (usually) one metallization mask layer. This means that the silicon foundry can have a small range of basic circuits available in which most of the mask layers are standard, maintaining the cost advantage of economies of scale.

As the customized layer of metallization is isolated from the rest of the circuit, apart from at the pre-determined contact points, there is flexibility in the organization of the array of gates. Three possibilities are illustrated in Figure 4.9. In the first approach, the so-called 'sea-of-gates', the greatest density of gates is offered. This advantage must be traded against the complexity of the design of the metallization layer and the restraints of the topology to avoid crossing of interconnect lines, and also to avoid open contact holes and so prevent spurious connections to the wrong gates. This requires very sophisticated routing software (the design tools are invariably automated) and is further complicated by the requirement for power and ground lines to the gates, although these may well be taken through on a different metallization layer, at the expense of a more complex fabrication process.

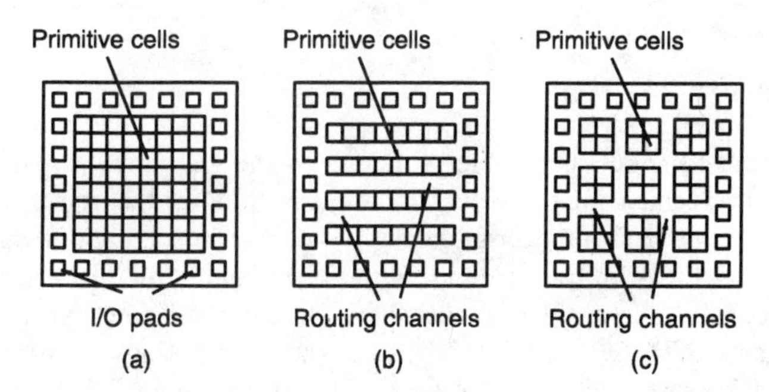

Figure 4.9 Possible gate array configurations: (a) sea of gates; (b) columns of gates; (c) clusters of gates

These problems are alleviated by the other two approaches illustrated in Figure 4.9. Here there are dedicated routing channels, often with pre-designed links. This greatly simplifies the routing algorithm and can provide a more flexible design approach, but at the cost of a lower density of gates on the chip (and hence greater expense of silicon) and the possibility of longer interconnects, increasing signal loading and reducing speed. These advantages and disadvantages are increased as one moves from the column arrangement (Figure 4.9b) to the cluster arrangement (Figure 4.9c).

4.4.2 Field programmable gate arrays

The alternative approach to the gate array realization is to employ the blowable fuse technology described earlier. In this way the gate array circuit can be manufactured in its entirety from a standard mask set and the customization comes later with the pattern of connections being made by the blowing of the fuses. With this approach it is difficult to use the 'sea-of-gates' type of array, as there is no space between the gate cells to arrange the interconnects. Very recently a three metal technology has become available to achieve this (see Chapter 7). The arrangement is usually of the column or cluster type of Figure 4.9. In addition, as there is potentially less flexibility in the design, the sizes of the routing channels are generally larger than in the mask programmed device, in order to provide sufficient routing combinations.

As with the PROM circuits described earlier in the chapters there is the possibility of 'single-shot' programming based on the miniature fuse, or the erasable version based on the floating gate MOS structure. However, the extra circuit area of the erasable version tends to make it prohibitively expensive compared with the fuse-based circuit. It can be cheaper to buy several 'single-shot' ICs than one reprogrammable version, and so this latter type is not currently very common.

4.4.3 Primitive cell design

We have talked about the general structure of the gate array circuit and the possible arrangement of the cells, and we now look at the content of these 'primitive' cells. In fact the term 'gate array' can be something of a misnomer. In some cases the primitives are indeed individual gates, but this is more often than not untrue. The problem is that there are many different types of gates, for example AND, OR, NAND, NOR, as well as flip–flops. In addition, the number of inputs to the combinational cells can be varied. As the gate array should have a reasonable selection of each type to maintain the flexibility and usefulness of the IC, the routing problem can become very difficult if a particular network of gates is required. This can lead to very long interconnection paths, poorer performance and very inefficient usage of the array, or even the inability to route a particular design on a particular IC.

There are two practical approaches to improve the flexibility of the array. The first is to have an array of transistors rather than gates, that is to take the problem one step further down the circuit hierarchy. So each primitive cell now contains a transistor or, in the case of CMOS, a pair of complementary transistors. The routing exercise also includes connecting up the transistors to form the required gates, as well as interconnections at the gate level of abstraction. In this way, the required gates can be formed in close proximity, and the interconnections between gates are very short and easy to route, leading to a much more efficient implementation.

This approach does not readily lend itself to the FPGA type of structure, owing to the high complexity of fuses and interconnects that would be required to maintain complete flexibility of design. Instead this transistor array approach is applicable only to mask programmable gate arrays. The alternative approach used by FPGAs is to make use of universal logic modules (ULMs), which are relatively small units of logic that can be configured to operate as different types of gates depending on the input conditions. There are a number of different realizations of ULM, one of which is particularly useful and flexible being based on a multiplexer circuit (Zhang, 1987). The primitive is a three-input circuit, one input being the select bit for one of the two other inputs to be directed to the output. The primary output is given by the Boolean expression $f = AB + \bar{A}C$, the gate level version being illustrated in Figure 4.10, along with the 'black-box' representation.

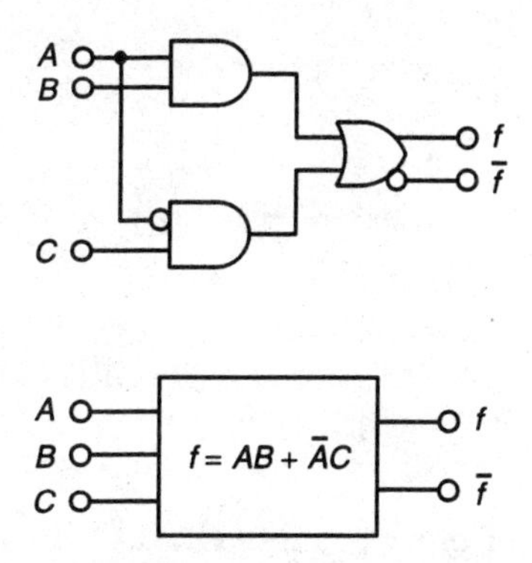

Figure 4.10 Multiplexer-based universal logic module

The ULM can now be configured into all the basic gate functions by control of the various input lines, as illustrated in Figure 4.11. In Figure 4.11a the two-input AND and NAND functions are generated by holding input C at logic 0. In this case the output function reduces to $f = AB$, the AND function between A and B, and the inverse output $\bar{f}$ is the NAND function. Figure 4.11b shows how the two-input OR and NOR functions are generated. In this case input B is held at logic 1, now $f = A + \bar{A}C = A + C$. Again the inverse output generates the NOR function. Figure 4.11c illustrates how the ULM can be used to extend the number of inputs to a gate. Here two ULMs are required to form a three-input AND/NAND. The technique can be extended to more inputs by continuing the cascade and multiple input OR/NORs can be formed in a similar fashion. Figure 4.11d illustrates how the EX-OR/EX-NOR functions can be generated. In this case $\overline{B}$ is input to C, so the primary output $f = AB + \overline{A}\,\overline{B}$ which is the EX-NOR function.

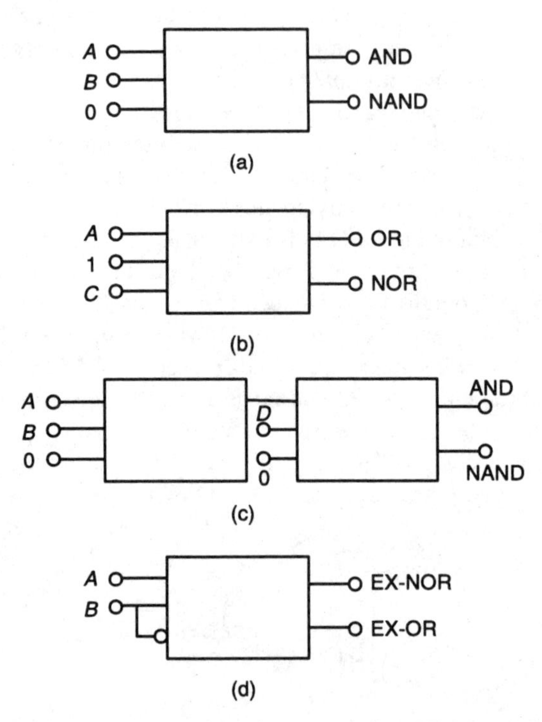

Figure 4.11 Implementation of basic gates using ULM: (a) 2-input AND/NAND; (b) 2-input OR/NOR; (c) 3-input AND/NAND; (d) 2-input EX-OR/EX-NOR

There are a number of other possible ULM configurations with a similar degree of flexibility. The multiplexer approach detailed here has a particular advantage in its circuit realization. It would appear from Figure 4.10 that the circuit is relatively complex, requiring several gates to generate the output functions. However, employing CMOS technology and using a transmission gate approach, the functions can be very easily generated, as illustrated in Figure 4.12. It can be seen that the ULM has no more complexity than an individual gate and hence occupies a minimal amount of silicon area.

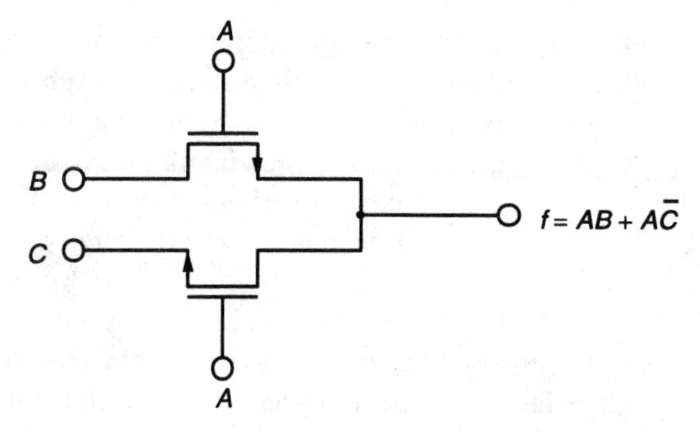

Figure 4.12 CMOS implementation of ULM

4.5 Standard cell

While modern gate arrays can now provide a reasonably good degree of flexibility and a way of rapidly prototyping and developing custom ICs, an even greater degree of flexibility and more efficient use of chip area can be obtained with only a small increase in design time and software complexity. This is the standard cell approach.

This process consists of a library of standard elements that the designer can make use of; schematics are available for all the standard elements, in a computer aided design (CAD) environment, in order to define the circuit. These cells may simply be individual gates, or may be larger blocks of circuitry such as adders, multiplexers, counters, shift registers, etc. Each cell is a self-contained design with defined positions of inputs and outputs via which connections may be made to other cells. However, different versions of the same function block may be available with different positions of I/O connections, different orientations of the cell, or even different size or aspect ratio of the cell. The size of the library and the number of different variants of the cell will depend on the manufacturing foundry that was responsible for development of the cells, their schematics and equivalent circuits (for simulation purposes, see Chapter 5).

Each cell also has a mask pattern associated with it, so that when the full circuit has been designed, the complete mask set will have been customized, as opposed to the mask programmed gate array circuits where only the interconnecting metallization layer is customized. This is the primary reason why the turn-around time to generate the complete circuits is longer than that of gate arrays, which have most, if not all, of their mask layers pre-defined.

The designer may have the capability of operating directly at the mask level, placing the cells where he or she wishes and wiring up the interconnections at the metallization layer. More usually, however, the designer will work at the schematic level, drawing the circuit in symbol and block form. The computer will then take the layout of the cells and their connectivity and use this information to optimize the position of the cells relative to each other and their interconnection, deriving the customized mask set automatically. As the place and route algorithm and optimization routines involved in such a process are extremely complex, the software can sometimes 'get it wrong', so careful checks have to be made that the circuit placement and connections are indeed correct. This is covered in more detail in Chapter 5. In addition, the designer can often edit the automated layout and so correct any computer errors or change aspects of the layout that may cause practical problems. As this requires a lot of experience on the designer's part, only skilled designers should go through this cycle of design.

To ease the optimization task and try to eliminate the designer's 'correction' cycle, standard cells are often arranged in a column layout with routing channels, very similar to the gate array architecture of Figure 4.9b. The algorithm still has the ability to place the cells in any order, so minimizing the connection lengths between cells and resulting in a more optimized solution than the gate array. It may even be that the size of the overall standard-cell IC is tailored to the size of the core circuitry (that is, not including I/O pads), where the overall size of the gate array IC is fixed. This approach greatly simplifies the place and routing algorithm, but will not lead to such an optimized circuit as the approach that has complete flexibility of placement.

4.6 Full-custom circuits

For the greatest flexibility of design and layout, leading to the optimal use of silicon area and the best performance in terms of speed and power, the designer must resort to the full-custom approach. Here there is complete freedom in the layout of the circuit, not only in the positioning of the gates and connection paths between them, but also in the size and shape of the circuit components, that is, much of the design is done at the mask or polygon level.

The task of a designer having to design an IC in its entirety 'from scratch' is of course a virtually impossible one, so even with the full-custom approach use is made of standard cells and blocks of circuitry. The difference in this case is that the designer has the ability to edit these cells to stretch the size or modify the topology, if this is required. There also exist a number of computer tools to aid the designer in this task. Based on a connection netlist or schematic layout, the software can optimize the interconnection and also recommend placement of cells. Recent developments in computer tools can also be used to highlight problems in the critical placement of important signal or clock lines to minimize cross-talk problems or possible bridging errors, as well as placement of accessible nodes to improve the testability of the circuit. Also the final circuit can be 'compacted' by the computer, based on the foundry design rules for the minimum spacings between the various mask layers, to minimize the overall circuit size.

The design task can still be a huge one and requires designers with some experience. Often the task is divided between a design team, requiring many man-months of work, and so is the most expensive in terms of development cost. However, the final circuit should be the optimal realization, and hence the cheapest to produce per circuit. The extra development cost will have to be recovered by this reduction in production cost, and this may not be possible unless a certain number of ICs, or the systems in which they fit, can be sold. So the full-custom approach is only viable for large production runs, not for small quantities of the custom design. The choices to be made and the economics involved were discussed in Chapter 1.

4.7 Analogue ASICs

The choices for custom design of analogue circuits and the associated software tools are much more limited than in the digital case. This is due largely to the greater diversity of analogue functions and there being no real analogue equivalent of the digital gate or ULM. There do exist a number of analogue cells, such as op-amps, comparators, oscillators, voltage references, etc., but even with these there can be a very large range in the specifications for the required cell. As well as these basic building blocks, the foundries often supply larger circuit blocks such as ADCs and DACs. There also exist other components such as resistors, capacitors, switches and inductors (for high-frequency circuits) which are not normally available in custom digital circuits.

The concept of a 'gate array' for analogue circuits does not therefore exist, and the only choices available to the designer are those of standard cell and full custom. The advantages and disadvantages of each are similar to those outlined above for the digital circuits.

4.8 Multi-chip modules

The designer has a choice to make in realizing a complete electronic system in terms of the level of integration of the circuits. At one extreme the system can be made up of many ICs, each of low integration level (gate packages etc.). These ICs are individually packaged and mounted on a printed circuit board. When ICs were first introduced and were only realizable in SSI or MSI, this was the only way in which a system was achievable. At the other extreme, the complete system can be fabricated on one substrate. This is becoming increasingly possible as improvements in process reliability make 'wafer scale integration' practical and also as mixed signal circuits are developed. The disadvantages of the first approach are cost, size, weight and reliability, although it is likely to be more testable and individual ICs can be replaced if a fault is detected. The second approach is generally cheaper, and has lower size, weight, power consumption, etc., but can suffer problems of testing and repair is impossible. Problems of yield exist in wafer scale integration. Even in the most reliable of processes it is almost certain that some defects will occur in ICs of this size. This drawback can be offset by the use of redundancy of critical elements and 'voting' procedures to ensure correct functioning of the system. The cost/size advantages have usually won out, and over the years the move has been to larger and larger levels of integration.

There is a further consideration to be made in system realization, namely how the various elements in the system are interconnected. Both the board and wafer approaches use the same basic interconnection line – a metal film on, or in, a dielectric medium, the difference being in the relative size and ratios of the metal dimensions. As operating speeds have moved into the high MHz frequency range, so transmission line theory has to be applied to these interconnections. The delay associated with a particular transmission system depends on such factors as the dimensions of the interconnecting line, the dielectric constant of the substrate, the capacitive loads being driven, etc., and is a very complicated function of all these parameters. A general rule-of-thumb is that, all other considerations being equal, the wider a line can be made, the lower is the delay. This only really holds for unloaded lines with an ideal driver; once practical loads and driver impedances are taken into account, the associated delay of the system can increase again with increasing line width. Typical curves are shown in Figure 4.13, which indicates that in practical systems there is a optimal line width to achieve minimum delay. While on-chip metal interconnections are useful for short interconnections, they can suffer from large delays with line lengths greater than several millimetres, which is another drawback with wafer scale integration, resulting in such circuits not being able to be used at high signal frequencies. In most realizations, the two approaches of on-chip connections and PCB connections lie at the two extremes of the upper curve in Figure 4.13.

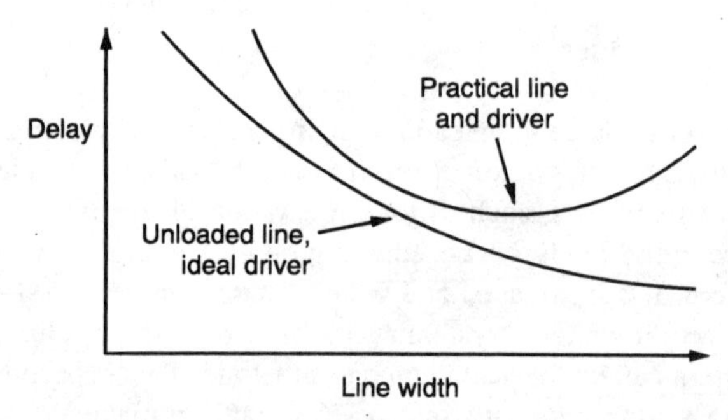

Figure 4.13 Typical delay–bandwidth characteristics for IC transmission lines

Recently, however, a technology has emerged that makes available the advantages of both these approaches, and that also has the flexibility of using the optimal line width to minimize the signal delays. The technology is termed multi-chip modules (MCMs) and consists of individual, unpackaged, medium to large scale ICs mounted on a silicon, or possibly other dielectric, substrate that contains the inter-chip interconnections. As these interconnections are not formed at the same time as the on-chip connections, they are not constrained to the same metallization thickness, so the thickness and widths of these interconnections can be optimized within the particular technology used to achieve the shortest delay properties and hence maximize the operating speeds. This approach has the advantages of still being relatively small and lightweight, no bulky printed circuit boards are required, while the hybrid approach eases the testing problem and allows the possibility of reworking faulty systems.

The choice of base substrate for the MCM depends on a number of factors. Prime among these is the complexity of the required routing interconnections between the ICs. It is highly unlikely that the topology is such that the routing can be done in a single layer with no cross-overs. Two or three layers of metallization may be required to solve the routing problem, so a series of metallizations with insulating dielectric layers and via holes is needed. It is possible to build such structures either on a silicon substrate with subsequent processing of the dielectric and metallization layers, or alternatively laminating a series of thicker dielectric layers together and dispensing with the solid silicon substrate. The 'silicon on silicon' approach, with the thinner metallization thicknesses, means that narrower lines can be fabricated, the density of interconnections is greater, and the overall module is smaller and likely to operate at greater speed. However, there is a practical limit to the number of interconnection layers that can be formed on the silicon substrate, typically two or three. With the laminated approach, the number of layers is virtually unlimited. As a result there is much more flexibility in the routing and a more optimized pattern formed, also providing a reduction in substrate area. A further advantage of the silicon approach is that it is possible to form passive components such as resistors and capacitors monolithically. This can be a particular advantage when forming mixed signal (analogue/digital) systems. In the original hybrid approach (on ceramic substrates), such components were either thick-film or discrete surface-mount components.

As mentioned previously, the discrete ICs that are to be mounted on the MCM are bare, unpackaged dice. On the I/O pads, which are normally where the connecting bond wires would be connected for normal packaging, there are small solder blobs deposited. The IC is then mounted upside down ('flip-chip') with the solder blobs in contact with the substrate pads. The chips are heated sufficiently to melt the solder and make both an electrical and a mechanical contact.

Reference

Zhang, C. (1987) An Investigation into the Realization and Testing of Universal Logic Primitive Gate Array, PhD Thesis, University of Bath.

Bibliography

E.D. Fabricius, *Introduction to VLSI Design,* McGraw-Hill, New York, 1990.

R.L. Geiger, P.E. Allen and N.R. Strader, *VLSI Design Techniques for Analog and Digital Circuits,* McGraw-Hill, New York, 1990.

P.J. Hicks, *Semi-Custom IC Design and VLSI,* Peter Peregrinus, London, 1983.

R.H. Katz, *Contemporary Logic Design,* Benjamin/Cummings, Redwood City, California, 1994.

M.J. Morant, *Integrated Circuit Design and Technology*, Chapman & Hall, London, 1990.

C.J. Savant, M.S. Roden and G.L. Carpenter, *Electronic Design, Circuits and Systems,* 2nd edn, Benjamin/Cummings, Redwood City, California, 1991.

M.J.S. Smith, *Application Specific Integrated Circuits,* Addison-Wesley, Reading, Massachusetts, 1995.

M. Sriram and S.M. Kang, *Physical Design for MCM,* Kluwer Academic Publishers, Boston, Massachusetts, 1994.

W. Wolf, *Modern VLSI Design – A CAD Based Approach,* Prentice-Hall International, Englewood Cliffs, New Jersey, 1993.

Questions

4.1. Construct the truth table of a 1-bit full adder to include an input carry bit C_{-1} in addition to the two summed bits A_0 and B_0. The output comprises two bits, S_0 and C_0. Directly from this truth table, construct the PLA implementation of the full adder circuit.

4.2. For the full adder of question 4.1, derive the minimized Boolean functions for S_0 and C_0 and construct a new PLA circuit to implement the functions. Comment on the results.

4.3. Using the multiplexer ULM circuit described in section 4.4.3, construct circuits to implement:
(a) a 4-input AND/NAND (minimum 3 ULMs);
(b) a 6-input OR/NOR (minimum 5 ULMs);
(c) a 1-bit full adder (minimum 5 ULMs).

4.4. Using a single ULM, implement:
(a) an R–S latch;
(b) a clocked D-type latch.
[*Hint*: in each case, feedback the Q output to one of the ULM inputs.]

5 CAD
How can we make the tasks possible?

5.1 Introduction

The task of design, fabrication and testing of even the simplest IC involves many operations and different stages of development. An error in any of these operations can render the final circuit inoperative, so careful checking of the design stages is therefore imperative. Many of the operations are repetitive, tedious and error prone, and are therefore much more suited to computers than humans. In addition, the sheer time that the operations would take to perform manually to produce an IC of greater complexity than SSI makes the wholly human-generated IC a virtual impossibility, even ignoring the economic disadvantages of such a route.

The use of computers is therefore essential. During the development of the various IC technologies and the increases in circuit complexity, more and more computer aids have been developed and made available to ease the burden of circuit design and to ensure, as far as possible, the elimination of errors. Computers can now be used, to a greater or lesser extent, in all the stages of IC realization as outlined in Chapter 1. This chapter will describe the various computer tools available to the IC designer, giving examples of state-of-the-art programs where appropriate.

5.1.1 Hierarchical levels

Before describing the specific computer tools, it is worth discussing briefly the various levels of hierarchy or abstraction at which the integrated circuit can be considered. This arises because the CAD tools largely operate at one particular level of abstraction, although application of the tools to mixed levels is becoming more common.

From a 'top-down' approach, these levels of hierarchy are:

System level. This may consist of one or more boards of ICs or an MCM system and its associated interconnections, forming a complete integrated system.

Chip level. This consists of a complete IC, and may be considered as a packaged device, with associated parasitic elements, or a complete, unmounted circuit die.

Sub-circuit or register transfer level (RTL). This consists of a significant piece of circuitry with a particular function. In the digital area, for example, it may be a multiplexer, adder stage, shift register, memory block, etc., while in the analogue area it may be a filter, oscillator, ADC/DAC, etc.

Gate or circuit level. This is the 'building block' level – individual gates in the digital circuits, op-amps, voltage references, switches, etc. in the analogue area.

Transistor level. Here we are at the individual component level, mainly transistors, but also including capacitors, resistors, transmission lines, etc.

Mask or layout level. The lowest level of abstraction is the mask level which contains the various polygonal shapes at different layers that are used in the actual IC fabrication and, when combined together, provide the exact topology of the circuit.

5.2 Circuit drawing

One of the first tasks of an IC designer is to sketch out a 'paper' design of part or all of the circuit. In the days before computer drawing aids were available, the designer had to have some draftsmanship skills in order to produce a clear drawing. The drawings were generally made in pencil so that subsequent alterations could be added to the drawing, redrawing the whole schematic was tedious and the whole process was error prone. In addition, the drawing was only useful as an illustration, and transfer to the actual circuit layout for construction was also a human task.

With the development of computer drawing aids and high-resolution colour displays, the task has been made infinitely easier. In addition, the computer package can use the circuit drawing to derive a 'netlist' which describes the topology and connectivity of the circuit elements. This can then be applied by other packages such as simulation, layout, routing, printed circuit board (PCB) and test pattern generation software. The whole process is referred to as 'schematic capture', as the circuit is represented in the schematic form of a drawing, but the information contained in the drawing is captured by the software for use in other design stages.

The basis of a schematic capture package is a library of circuit elements that the designer can call up and place in the circuit drawing. The prime function, in terms of the drawing exercise, is to eliminate the need to redraw the symbol for, say, a NAND gate every time one is required in the circuit. But the library element usually contains much more than just an outline drawing of the part – it will have some reference attached to it that will indicate a physical device. The form of the device will vary, depending on the particular schematic capture package and the other packages it refers to; for example, some products are aimed at using standard gate packages for realization of the circuit on a PCB. In this case, for instance, if the designer uses a two-input NAND gate from the library, the reference will be to one of the four components contained in the 7400 IC package. When the netlist is subsequently used to generate the PCB artwork, a 7400 package is placed and connected to the rest of the circuit. If other identical parts are placed, the software will use other parts from the same IC package until all four are utilized. If, however, the schematic package is to be used in a standard-cell technology, the reference in the library will be to the appropriate cell that can be placed and routed in the layout routine.

So from the library, or set of libraries, information can be obtained on the outline drawing, the reference to the physical device or part, its possible orientations in both drawing and circuit placement, equivalent circuits or mathematical models for simulation purposes and even data on its testability or appropriate test patterns.

In general, drawing packages have three main elements: components, wires and text. Input and output connections are generally considered as components; in board layout they may be edge connectors, for example, whereas in chip layout they may be bonding pads. The wires represent connections between components, along which the signals are transmitted. A group of connected wires, having a common signal associated with them, is referred to as a 'net' (hence 'netlist' for the complete circuit description). Some drawing packages automatically connect wires to form nets whenever a T-junction is formed, other packages require the designer to specify each and every connection between wires. Most packages have error checking routines to warn of unconnected wires or nets, where two component outputs are driving the same net and other possible drawing errors. The text facility can be used simply for labelling or adding information to clarify the circuit drawing, but can also play a more important role in assigning references and parameters to components and naming nets for reference in other packages.

All schematic capture packages have the ability to pick and place library symbols, and move, delete and copy components as required. They rely heavily on the graphics facilities of the computer. Most, if not all, of the operations can be done using a mouse, although keyboard operations are usually available as an alternative. Most now run under a 'windows' type of environment and much use is made of icons, all of these features making the drawing operations simpler to learn and execute.

The exact form of the computer file under which the drawing is stored will depend on the particular package and even the make of computer system that the package is running on. It may not be possible to accommodate large designs on a single drawing 'sheet', and if the overall design extends over more than one sheet, some form of reference between the common nets connecting the sheets must be established. The drawing may also be hierarchical, whereby a block of circuitry may be condensed into a 'black box', usually represented by a single polygon with the requisite input and output connections for representation at a higher level of drawing. It is usual that the designer can move between different levels of drawing without having to exit and reload designs. The various levels of hierarchy may not necessarily correspond with the levels of abstraction referred to in the previous section, but may simply be a way of including large amounts of (say) gate level circuitry on to a single design sheet. Such a hierarchy may extend over several levels. When the overall netlist is generated for simulation purposes (for example), the design is then 'flattened out' by the software so that all of the black boxes are expanded and the circuit can be operated on as a whole.

The form of the file in which the drawing information is saved is not usually in an acceptable form for the other software tools that will make use of the information in this file. The drawing file, as it is associated with a graphical representation of the circuit, is likely to have graphical references which may well be specific to the computer hardware and which are irrelevant to the connectivity of the circuit. It is almost inevitable that the drawing file will have to be converted at least into the form of a netlist that can be used by other design tools, and sometimes, additionally, into a textual description of the connectivity, or even a parts list, which is readable by the human designer, for reference and/or error checking.

There are a number of forms for describing a circuit in a text, other than machine-dependent, graphical format. Some of these are specific to particular software packages but there are also a few standard formats. Having these standards, and the computer tools to convert from the particular schematic capture format of the circuit description, enables portability between different hardware platforms and integration of different products into an overall design system to be achieved.

Such netlist standards include EDIF (Electronic Design Interchange Format) and SPICE (see section 5.3).

5.3 Simulation

Before software tools were developed, the verification of a particular circuit design could only be achieved by constructing a prototype circuit. While the designer could use standard digital and analogue techniques to design the circuit on paper, it was almost impossible to determine whether the circuit would perform as expected in practice. These 'breadboarded' prototype circuits were constructed using discrete components such as individual logic gates in the case of a digital design, or transistors, resistors, capacitors, etc. for analogue circuits. When built, the circuit would be thoroughly tested and design modifications made on the basis of these tests. The next version of the prototype was then constructed and the process repeated. Such a technique was very time consuming and expensive, and resulted in very long development times for a new product. There was a further complication if the product was to be in the form of a new IC, as the translation from breadboard to chip often resulted in a different performance owing to different device parameters, different parasitics associated with the circuit and errors in forming the fabrication mask set. Redesign and rework of the mask set added further delay and expense to the product development.

The solution to the problem is to simulate the operation of the circuit before construction of a prototype. Any problems highlighted by the simulation can mean a much quicker design iteration and, if the simulation is sufficiently accurate and the designer checks the design thoroughly, this should result in a 'right-first-time' design. The advantages in time and expense over the old process are obvious.

The quality of a circuit simulation is critically dependent on two parameters. First is the simulation routine itself, that is how the mathematical relationships between the signals are implemented and how closely they match the true relationships in the circuit; and second how accurately the components that comprise the circuit are modelled.

Simulations can be done at any of the levels described in section 5.1. Digital simulations are mostly done at the gate level, while analogue simulations are done at the transistor level. True mixed signal simulations are very difficult to perform, partly because of this different level of hierarchy, and are therefore rare.

5.3.1 Digital simulation

In principle, digital simulation is very straightforward: the signals have only two possible states, 1 and 0, and the outputs from the gates are well-defined functions of the inputs. Hence very simple routines can be constructed to predict the output of a circuit based on the conditions of the inputs. Such a simulation is termed a switch level simulation and can be used to verify the basic functionality of a digital circuit to check that no errors have been made in the gate level design. However, the switch level simulation will not predict whether a particular design will work in practice. The main problem is the fact that there will be a delay associated with a gate, in that the output will not change immediately a change in the input signals occurs. This gate delay may also be dependent on the loading on the gate. The effect of gate delay can result in short erroneous signals on certain nets as one input signal to a gate may change before another, where the switch level model predicts they change at the same time. The cumulative effect of the gate delays will put a limit on the overall speed at which the circuit can operate.

As well as including time delay information in the model, other features of more advanced gate level simulation include the addition of other signal levels or states. As well as the standard 1, 0 and don't care, which can be implemented at the switch level, unknown states can be included. In general, all nets are set at an unknown state at the start of a simulation until such time as the effects of the inputs change them into a known state. This can indicate redundant nets, or be used in fault analysis to identify unexercised nets, as well as checking that the basic functionality is correct. Other signal levels include 'high impedance' where a net is effectively open circuit, for use in tri-state I/O for example.

Digital simulations are usually done in the time domain, that is a simulation of the values of the nets is carried out as a function of time. This simulation may be done by taking equal steps in time and analysing the value of each net after that time period. However, with this approach a small enough time step must be taken to ensure that no short time period transitions are overlooked. As there may subsequently be comparatively long time periods when no signal changes, the analysis will be repeated many times for no essential reason, thus wasting calculation time. Instead, the 'event driven' approach is nearly always substituted. In this case, any change in a signal triggers a re-analysis of the circuit to see what effect the change has on connected parts of the circuit. Once the effects have been analysed, the time is advanced until any other signal change is identified; so no unnecessary calculations are made during periods of signal inactivity.

The usual output from a digital simulator is a graphical plot, or corresponding text list, of the signal values on the nets as a function of time. While the true signal voltage varies within the limits of the acceptable logic levels and the transitions between levels take a finite time and are usually of an exponential shape, the plots are at discrete levels and the transitions usually shown as instantaneous. The position of the transition, however, is based on the time at which the signal voltage passes the particular logic threshold.

There are many tens of commercial digital simulators available, most associated with a particular vendor and usually integrated with a suite of software packages relevant to a particular product. The component libraries are usually modelled on a particular fabrication process, though most can be edited by the user to alter parameters such as gate delay.

5.3.2 Analogue simulation

Analogue simulation is potentially much more involved than digital simulation; this is partly due to the greater number of variables and parameters that can be of interest. For example, both time and frequency domain performance can be of interest, and variables can include not only voltages and currents but also noise, gain, input and output impedances. In addition, component tolerances are of more importance, so sensitivity analysis can be considered and the circuits can also be linear and non-linear in nature, requiring different simulation approaches in each case. Another difference is that it is very difficult to treat analogue simulation in the time discrete way in which event driven digital analysis is performed. The simulation must be done in a time continuous way, or at least in a series of discrete time (or frequency) steps of sufficiently small value to model the circuit performance accurately.

Having said that, there is an almost 'standard' approach to analogue simulation which is a time domain, nodal analysis based on the circuit connectivity, as defined by the netlist, and transistor level models of the components of the circuit. A simulator that is based on this approach and that is almost universally used in one form or another is SPICE. Because of the widespread use of this package, it will be described in some detail in section 5.3.3.

The quasi-time-continuous nodal analysis applied to the transistor level of circuit abstraction can provide solutions to most of the analogue simulation problems outlined above. Being a nodal analysis it can readily provide a solution to the node voltages and branch currents at each time step, and impedances can be derived from the voltage and current solutions. Frequency domain results can be derived from a Fourier transformation of the time domain data; this is usually achieved via the discrete mathematical routine known as the Fast Fourier Transform (FFT). Noise analysis can also be incorporated by the inclusion of noise voltage or current sources at the appropriate points. Whether the analysis is linear or non-linear in nature depends on the particular element models and whether the components of the model have other dependencies on signal levels, for example. Likewise, temperature effects can be incorporated by modelling the components with accurate thermal dependencies.

The fidelity of these simulations in predicting the true behaviour of a real circuit depends again on the two factors of the mathematical algorithm and the quality of the models – both can be extended to improve fidelity. The basic nodal analysis is in itself rigorous, but can suffer from inaccuracies in its computer implementation, as the solving routine can become mathematically unstable or insoluble. The routine should produce warnings when such situations occur. Inaccuracies can also occur as a result of the machine resolution of the floating-point calculator. Both of these sources of error are exacerbated by variables, such as voltage and current, which differ by many orders of magnitude. Another consideration is the time step between each calculation. This is usually specified by the user, but some sophisticated routines can detect when variables are changing quickly compared with the size of the time step. This is another potential source of error or instability, so the routine will automatically decrease the time step to offset the problem. There is generally some minimum time step specified that the automatic routine cannot exceed.

The models for devices such as transistors can be made extremely complex (see section 5.3.3), but should be based on the physical structure and behaviour of the device. Much work is devoted to the modelling of semiconductor devices to optimize their linear and non-linear behaviour.

The cost of these accuracy enhancements, both in the modelling and the analysis, is one of computing time. This means that if circuits are analysed in such a way at the transistor level, the designer is limited in the total number of devices in the circuit unless the simulator run times are to become prohibitively long. Typically this is of the order of hundreds of devices before run times in hours are encountered. This means that the simulation of LSI circuits and above are not directly practical.

The long simulation times associated with an analysis at circuit/gate level for both analogue and digital circuits have led to the development of more functionally based, hierarchical techniques for modelling of circuits, termed hardware description languages. These are described in section 5.4, but before we leave the section on simulation, the 'standard' approach to analogue simulation, SPICE, is described.

5.3.3 SPICE

SPICE (Simulation Program with Integrated Circuit Emphasis) was developed at the University of California in the mid 1970s and subsequently released as a public domain software package. The early versions used purely a text input (the circuit netlist and other commands and parameters, see example below) and generated a text output (voltages and currents at each time, or other parameter, step). As these text forms are not the most friendly for a designer to work with, different pre- and post-processors have been developed to interface with the main simulation engine. The simulator still requires the input in the correct text format, although most schematic capture packages can generate a circuit netlist in the SPICE format. The conversion of text output data into a suitable graph display is straightforward.

SPICE operates at the transistor level of abstraction and can perform d.c. and a.c. analysis; d.c. analyses include operating point and swept d.c. while a.c. analyses include small signal (linear), non-linear and noise simulations. Transient analysis can also be performed. For most analyses, the program has initially to determine the d.c. operating point, which it does based on the value of any d.c. sources in the circuit. For all other nodes it assigns an initial guess to the value of the d.c. voltage and then iteratively solves the network to derive the correct operating point voltages. The closer the initial guesses are to the actual voltages, the faster the routine will iterate to the correct value. The convergence to a solution is a critical aspect of the operation of SPICE and sometimes the routine will not converge, so it is possible to preset the initial node voltages to aid or even permit convergence.

Once the operating point is determined, the program can continue to the other analyses, again based on an iterative solution to the network equations. For d.c. swept analysis, the appropriate input voltage source is stepped through the specified values and the circuit re-analysed at each step to provide a transfer curve. Single-frequency or swept a.c. analyses can be performed, or a transient analysis as a function of time. In each case the parameter is stepped through all those specified by the user and the network iteratively re-analysed. Temperature and sensitivity analyses can also be covered, based on the parameters of the individual element models.

An example circuit, a CMOS differential amplifier, is shown in Figure 5.1 and the corresponding SPICE input file, complete with the circuit netlist, control and input/output parameters is given below.

```
* FILE: MCDIFF.SP
CMOS DIFFERENTIAL AMPLIFIER
.OPTIONS SCALE=1E-6 WL
VIN 7 0 0 AC 1
.AC DEC 10 20K 500MEG
.PRINT AC VDB(5) VP(5)
M1 4 0 6 6 MN 100 10
M2 5 7 6 6 MN 100 10
M3 4 4 1 1 MP 60 10
M4 5 4 1 1 MP 60 10
M5 6 3 2 2 MN500 10
VDD 1 0 5
VSS 2 0 -5
VGG 3 0 -3
RIN 7 0 1
.MODEL MN NMOS LEVEL=5 VT=1 UB=700 FRC=0.05 DNB=1.6E16
+ XJ=1.2 LATD=0.7 PHI=1.2 TOX=800
.MODEL MP PMOS LEVEL=5 VT=-1 UB=245 FRC=0.25 TOX=800
+ DNB=1.3E15 XJ=1.2 LATD=0.9 PHI=0.5
.END
```

The circuit topology is defined in terms of node numbers. These are shown in Figure 5.1 by numbers within circles for easy reference to the netlist above.

The first line in the file is a statement of the file name. However, it is preceded with an asterisk which means the line is simply a comment and is ignored by the simulator. All such SPICE input files have the file extension `.SP`. The second line is the title line which is a brief description of the circuit. The program automatically takes the first valid line of the file to be the title line.

The next section, defined by the `.OPTIONS` command, allows the user to set a series of input control options for the simulation. The `SCALE` parameter is a size multiplier that applies to the active elements, in this case the MOSFETs. By setting them to 1E−6 indicates that the dimensions in the file are in microns. The `WL` parameter inverts the order of the width and length dimensions in the MOSFETs specification lines. All the options should be

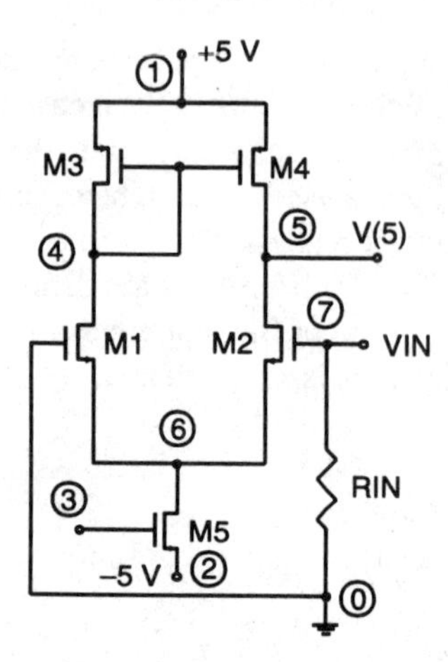

Figure 5.1 CMOS differential amplifier

contained within one program line (a program line can be continued over more than one text line by placing a + at the start of the next line, as in the .MODEL command lines above).

The next section defines the input stimuli to the circuit and the type of analyses required. In this case there is only one input signal, VIN, and the analysis required is swept frequency AC. The .AC command is followed by a three letter parameter. In this case it is DEC indicating that the step variation is a decade function. Other possibilities include LIN for a linear variation, OCT for octave variation and POI for a list of frequency points. The next number is the number of frequency points at which the analysis is performed, followed by the start and stop frequencies. The input voltage stimulus is defined by a name (VIN, the first of the three letters defines the stimulus as a voltage source). The next two numbers specify the node numbers across which the source is connected, in this case between node 7 and node 0. By default, node 0 is always the ground node in SPICE. The third number is the d.c. offset value, in this case 0. The source is then specified as an a.c. source, the final number being the magnitude of the voltage source in volts.

The next command line, .PRINT, specifies the output format. In this case the output consists of a list of the values of the a.c. voltage in dB and the associated phase of the signal at node 5 at each frequency point.

The next block consists of the circuit definition in a netlist format. Each line refers to a component and has the component name, nodal connection and value and/or reference information. For example, the first of these lines refers to transistor M1. The next four numbers indicate the node to which the drain, gate, source and bulk connections are made. The next parameter is a reference to a generic transistor model specified further down and the remaining two numbers are the width and length of the transistor gate structures. These dimensions are in microns as a result of the SCALE command earlier in the file, the normal default dimensions would be metres.

The supply voltages are also specified in this netlist. The default type for a voltage source is d.c., so the three numbers after the element name are the two node numbers across which the source is placed; note that they are all referenced to ground, followed by the voltage value.

The remaining statements, preceded by the command .MODEL, provide more detail about the MOS transistors. There are two statement lines, one for the NMOS device referenced by the label MN and one for the PMOS device referenced by MP. The LEVEL command references the complexity of the transistor model used. Depending on the version of SPICE used, there are up to tens of different levels available, some being universal models while some are specific to certain technologies or manufacturers. As stated previously, the more complex the model, the more accurate the final simulation is likely to be – but at the cost of computing time. In this example the model used is the level 5 model based on the work of Huang and Taylor (1975). The other parameters specified in the command line are given below with their dimensions:

VT – Threshold voltage (V)
UB – Low field bulk mobility ($cm^2\ V^{-1}\ s^{-1}$)
FRC – Field reduction coefficient (Å s cm^{-2})
DNB – Surface doping density (cm^{-3})
XJ – Junction depth (μm)
LATD – Lateral diffusion on each side (μm)
PHI – Built-in potential (V)
TOX – Oxide thickness (Å)

There are a number of other parameters associated with each model level, and these have default values that are applied if the parameter is not specified in the command line. For the level 5 MOSFET model, the parameters VT, TOX, UB, FRC and DNB must be specified, all others being optional.

The final line in the input file must be .END to inform the program that this is the end of the file.

This example has given just a flavour of the typical outlay of a SPICE file and the sort of parameters and models involved. There are many more options and specifications available than could be discussed. Different versions have different numbers of available options. Many CAD companies have developed their own versions of SPICE, the most commonly used ones (in ascending order of complexity) being SPICE2 (from University of California, Berkeley), pSPICE and HSPICE.

5.4 Hardware description languages

Most simulators operate at only one level of abstraction, for example gate level for digital and transistor level for analogue. Although they usually have the capability of defining larger blocks of circuitry in terms of sub-circuits or macros, these are more often than not 'flattened out' to the lower abstraction level before the simulation commences. A few simulators are truly 'mixed-mode' in terms of handling different levels of abstraction (not to be confused with the term 'mixed-signal' which is now more widely used to indicate mixed analogue–digital ICs). The problem with single mode simulators is the simulation time required to analyse circuits of greater complexity than MSI. A development that has grown in importance, mainly from digital simulators, is the use of high-level computer languages to perform the twin operation of describing the connectivity of a circuit and the functionality of the connected blocks, and this therefore can be used directly or indirectly in a simulation of the circuit performance. As the input and output signals are an abstract description rather than a voltage or current, and the relationship between input and output for a given block is a purely functional one, these languages can easily handle a mixed-mode description of the circuit. As such languages are used to represent the IC components they are termed hardware description languages (HDL).

A number of HDL tools have become available over the last decade or so, ranging in complexity from basic switch level simulators up to very comprehensive packages. The operation and typical structure of such tools will be illustrated below by means of some examples. A 'standard' has emerged recently that is becoming more available and accepted by the IC community. This was developed to cope with the new technologies and higher speed circuits being made available and so is termed VHSIC HDL or VHDL (VHSIC stands for very high speed integrated circuit). As this represents the future of HDL, it is described in some detail in section 5.4.2.

5.4.1 HDL examples

Prior to the general acceptance of VHDL as a standard HDL, CAD vendors tended to develop HDL tools that were specific to a particular suite of software, and so a number of different packages exist. To illustrate the structure and flexibility of these languages we will look at a couple of examples taken from the ES2 SOLO 1400 package, as this software was used in the IC design example of Appendix 2.

The SOLO 1400 HDL is termed MODEL and is a textual description of the connectivity and functionality of a digital circuit. It can be compiled from the circuit drawing generated by the schematic capture program, DRAFT, or generated by the user to create new functional blocks. Illustrations of both approaches will now be given.

The first example is of a two-to-four line decoder, illustrated in Figure 5.2. This circuit is designed to be used as a sub-circuit in a larger design, and a symbol that represents this sub-circuit function can be generated under DRAFT. The circuit has three inputs, which include an enable line that should be held low for the main circuit to function correctly. The other two inputs are the selection lines; the four possible combinations of these two inputs select one of the four outputs to go low while the other three output lines remain high. The circuit uses the library elements for the inverter and NAND gates. These will have corresponding layout definitions when the standard cell IC is compiled, as well as simulation information covering gate delays and fan-out limitations.

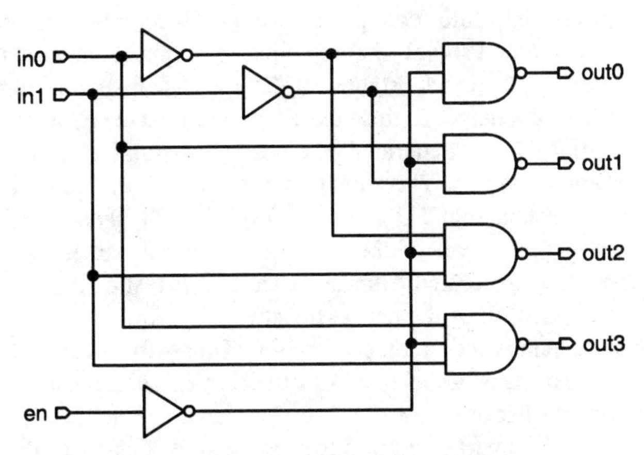

Figure 5.2 2-to-4 decoder circuit

The generated HDL file for this circuit is given below:

```
{ Design generated from design file dec2to4 }

Part dec2to4 [in1, in0 en]→out3, out2, out1, out0

  Signal s1a1x2, s1c2x1, s1c1x1

  {Output from sheet s1}

  nand [s1c1x1, s1a1x2, s1c2x1]→out0 : s1c3
  nand [in0, s1a1x2, s1c2x1]→out1 : s1c3y2
  nand [s1c1x1, s1a1x2, in1]→out2 : s1b3
  nand [in0, s1a1x2, in1]→out3 : s1b3y2
  not [in0]→s1c1x1 : s1d1
```

```
    not [in1]→s1c2x1 : s1c2
    not [en]→s1a1x2 : s1a1

End

External Signal in1, in0, en
External Signal out3, out2, out1, out0
dec2to4 [in1, in0, en]→out3, out2, out1, out0 : dec2to4

End of File
```

Lines enclosed in curly brackets are comments which are ignored by subsequent programs that use the HDL file. All the signals (equivalent to nets) must be named. These fall into two groups: internal signals, declared by the `Signal` statement, and signals that are input or output to the circuit, declared by the `External Signal` statement. The user can specify names for all the signals on the schematic. In this case only the external signals were named, so when the HDL file is generated, the software supplies labels for all the unnamed (internal) signals. The signal names are determined by the position of the net on the schematic. The drawing is on a single sheet, labelled `s1`, so this comprises the first part of the each of the signal names (a comment is made that the file is generated from this sheet). The rest of the name is a grid reference to the position of the net on the sheet.

The first statement is the `Part` declaration. This is followed by the name of the part, any parameters associated with the part in parentheses (in this case there are none), the list of inputs in square brackets, and the list of outputs following the arrow. The part definition follows, which consists of the list of signals and the connectivity of the components that make up the part and is terminated in the `End` statement. If there is more than one part in the overall circuit, the next part is defined. Once a part has been defined, it can be used (instanced) in further part statements. The components used in this part definition are basic gates available as library elements. If a component is used more than once in a part definition, it must have a unique label to differentiate it from the other components. This label could be provided by the user at the schematic stage; if not, a default label is given, based on the grid reference of the part on the sheet. Hence the netlist statements consist of the component type, inputs in square brackets, outputs following the arrow and the component label following the colon.

Once all the parts have been defined, the rest of the file is a circuit statement, in the same form as the component statements above. Indeed, the circuit is now defined as a new component that can be used in other circuits; the schematic capture program can also generate a circuit symbol for the decoder component, once the input and output signals are specified.

If the file defines the final, complete, chip-level circuit, the HDL file can also include definitions for the bonding pads. As stated previously, these are treated as separate library components with suitable models for the sake of simulation as well as chip layout information.

The example described here is a relatively simple one and is only given to indicate the type of structure used in HDL. It does not really illustrate the flexibility of describing the circuits in terms of a high-level language, being little more than a netlist file. The next example gives more insight into the power of HDL specification in the parameterizing of a component. The circuit is a pseudo-random binary sequence (PRBS) generator which consists of a shift register, from which 'taps' are taken from certain flip–flop outputs; the taps are EX-NOR'd and fed back to the input of the shift register. By altering the position of the taps, different sequence lengths can be generated. To realize a flexible circuit,which can be easily altered by specifying different parameters, use is made of a variable length shift register. The programmable PRBS generator is illustrated in Figure 5.3, and the HDL code is given below.

```
{ Design generated from design file prbs }

Include "synclib.inc"

Part vlsr (n) [ck,d,r]→q

  Integer i
  Signal qi(1:n+1)
  If n=0 Then

    d→q

 Else

    d→qi(1)
    For i=1:n Cycle

      dff rn [ck,qi(i),r]→qi(i+1) , - -

    Repeat
    qi(i+1)→q

  Endif

End

Part prbs [ck,r]→q
```

```
    Signal s1b2x1,s1c1x1
    { Output from sheet s1 }
    vlsr(3) [ck,s1c1x1,r]→s1b2x1 : s1b2
    vlsr(4) [ck,s1b2x1,r]→q : s1b3
    eqv [s1b2x1,q]→s1c1x1 : s1c2

End

External Signal ck, r
External Signal q

prbs [ck,r]→: prbs

End Of File
```

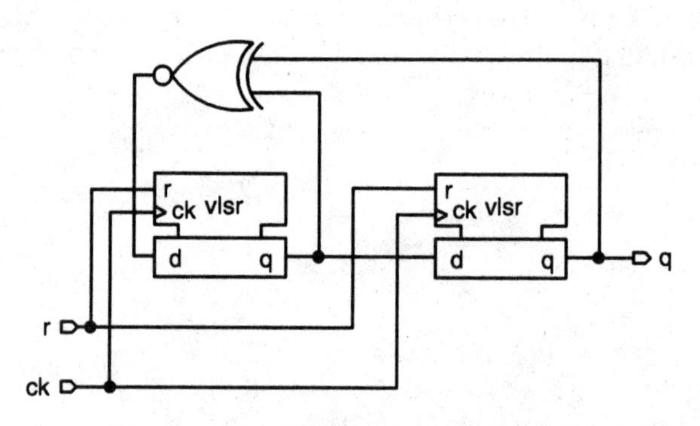

Figure 5.3 PRBS generator circuit

The first valid line of the file is to include access to a library of synchronous sequential gates; it is in this library that data on the D-type flip–flops resides. There are two parts defined: the variable-length shift register (`vlsr`), followed by the `prbs` circuit itself, which instances the shift register after it has been defined.

Looking at the definition of the `vlsr` part, in the `Part` statement there is a parameter specified in parentheses. This is the length or number of flip–flop stages and is given the variable `n`. The definition of the part contains features found in other high-level languages, such as `If...Then...Else` and `For...Repeat` constructions. The `If` condition is used to check for the special case when the shift register has length of 0, which would be an allowable parameter for `n` but which effectively means there are no flip–flops, that is the part has no circuitry! In this special case the input is passed straight to the output. The `For...Repeat` loop constructs the hardware description of the shift register, cascading the D-types `n` times and connecting the output of one to the input of the next. The output of the final stage becomes the output of the part, `q`.

The length parameters for the two `vlsrs` that make up the `prbs` circuit are defined in the schematic capture process; in this case they are `3` and `4` and these were transferred into the `prbs Part` definition when the HDL code was generated. Note that the exclusive NOR gate used in the circuit has the label `eqv` in the SOLO 1400 system.

5.4.2 VHDL

When hardware description languages were first developed, the software firms tended to develop their own forms to fit in with the other CAD tools that they marketed. However, the development of a standard version was desirable for portability between systems and VHDL was constructed with the object of being sufficiently flexible to interface with other products such as proprietary schematic capture tools and simulators, but with the emphasis on being able to handle the very high-speed new technologies that were emerging in the 1980s. The standard was adopted by the IEEE as number 1076 in 1987.

A full description of the structure of VHDL is beyond the scope of this text, but there are several textbooks that are solely devoted to this topic. This section contains a brief description of the basics, including the structure of the high-level language, the modelling of components and an example of a VHDL file.

VHDL consists of a text description that is both human and machine readable. The description of a component or circuit is a behaviourial one in a series of statements similar to those in other high-level languages (that is, the same sort of constructions in the SOLO 1400 HDL illustrated in the previous section). Complete systems are specified by a hierarchical arrangement of the interconnected components. VHDL is strictly for digital circuits, for although analogue HDLs exist, there is as yet no standard available; however a standard analogue version of VHDL is likely to be adopted soon.

VHDL can include language descriptions of structure and also at the register transfer level as well as behavioural circuit descriptions. This combination gives a very flexible approach and allows the verification of system design at different levels of abstraction.

A VHDL source file contains either a behavioural model, or a package, or both. A package contains constants, types and sub-programs used by one or more behavioural models. Both models and packages have a declaration and a main body. In the case of the model declaration or entity declaration, this contains a name that corresponds to a component symbol and a list of input/output ports corresponding to the component's pins. It may also contain parameters that define different attributes the model may possess. The main body of the model defines the architecture or behavioural operation of the component in terms of concurrent processes and via the high-level language constructs. Packages are not associated with a behavioural model, but are used to define sub-programs, constants or types separately from a model. Of course, a VHDL source file containing only packages cannot be simulated in isolation.

Objects are fundamental to VHDL files in that they contain the values within models. Each object has a type and a class. The type indicates the type of data it contains and the class defines what can be done with the data. Data types are further subdivided into pre-defined and user-defined. Pre-defined data types are: Boolean, representing one logical bit and having values true and false; integer, represented by a 32-bit word; bits, which can be 1, 0, unknown or high impedance; characters, having values of the ASCII character set; time, used in delay specifications; and text, having string values. User-defined data types include such things as enumeration types, having a single calculated value, and arrays or records, having more than one element. Object classes are constant, variable and signal. Constants have a single declared value, variables can have a value that changes with assignment, and signals are functions with time which are in general restricted only to bit types.

There are a whole series of operators that can be used to manipulate the data types. These include arithmetic operators such as addition, subtraction, multiplication, etc.; relational operators such as equals, greater and less than, etc.; and logical operators such as AND, OR, NOT, etc. There are restrictions as to the data types that can be operated on and the type of result. For example, relational operators can operate on any types, provided the left-hand side type corresponds with the right-hand side type in the operational equation, but the result will always be a Boolean (true or false).

There are a large number of sequential and concurrent statements that often appear in other high-level languages. Such statements include if...then....else constructions, loop, wait, case (conditional execution) as sequential statements and block, process, begin...end as concurrent statements.

Sub-programs, which may be defined within a package, consist of either a procedure or a function. A procedure may modify its arguments, but returns no value. A function may not modify its arguments and returns a single value.

As an example of a VHDL file, we shall look at the description of a half adder circuit as illustrated in Figure 5.4. The VHDL source file below has behavioural model descriptions for the three different types of gate, followed by a structural model for the complete half adder circuit.

```
- - Inverter Model
- - Port Definitions
ENTITY inverter_gate;

  PORT (A: IN BIT; Z: OUT BIT);

END inverter_gate;
```

```
- - Behaviour Definition
ARCHITECTURE behavioral OF inverter_gate IS
BEGIN

  Z = NOT A AFTER 10 ns;

END behavioral;

- - AND Gate Model
- - Port Definitions
ENTITY and_gate;

  PORT (A,B: IN BIT; Z: OUT BIT);

END and_gate;


- - Behaviour Definition
ARCHITECTURE behavioral OF and_gate IS
BEGIN

  Z <= A AND B AFTER 10 ns;

END behavioral;

- - OR Gate Model
 - - Port Definitions
ENTITY or_gate;

  PORT (A,B: IN BIT; Z: OUT BIT);

END or_gate;

- - Behaviour Definition
ARCHITECTURE behavioral OF or_gate IS
BEGIN

  Z = A OR B AFTER 10 ns;

END behavioral;

- - Half Adder Model
- - Port Definitions
ENTITY half_adder;
```

```
  PORT (A,B: INPUT; Sum, Carry: OUTPUT);

END half_adder;

- - Structure Definition
ARCHITECTURE structural OF half_adder IS

  - - List of Components Used
  COMPONENT inverter_gate
    PORT (A: IN BIT; Z: OUT BIT);
  END COMPONENT;
  COMPONENT and_gate
    PORT (A, B: IN BIT; Z: OUT BIT);
  END COMPONENT;
  COMPONENT or_gate
    PORT (A, B: IN BIT; Z: OUT BIT);
  END COMPONENT;
  - - Internal Signal Lines
  SIGNAL n1,n2,n3,n4: BIT;

BEGIN

  - - Mapping of Gates and Signals to Define Circuit
  g1: inverter_gate PORT MAP (A, n1);
  g2: inverter_gate PORT MAP (A, n2);
  g3: and_gate PORT MAP (B, n1, n3);
  g4: and_gate PORT MAP (A, n2, n4);
  g5: and_gate PORT MAP (A, B, Carry);
  g6: or_gate PORT MAP (n3, n4, Sum);

END structural;
```

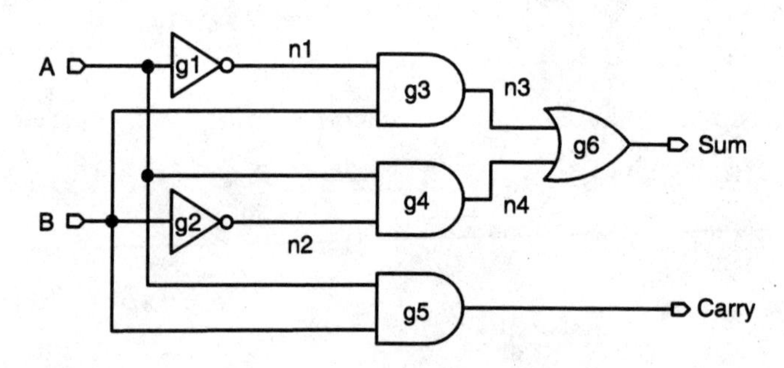

Figure 5.4 Half adder circuit

Gate delays of 10 ns are associated with each of the components; the lines preceded by the double dashes are comments and are ignored by subsequent software packages that use the source file.

5.5 Layout

In order to fabricate an integrated circuit, the series of mask patterns that are used to define the various layers must be defined. These patterns may be implemented as actual photolithographic masks or used as data for an electron beam 'direct-write' system (see Appendix 1). The masks are a series of overlaid patterns, mostly polygonal, often simply a series of rectangles. In the early days of IC manufacture these layout drawings were done by hand, an operation that was very intensive, time consuming and error prone. It was a part of the overall process that benefited most from computer implementation and being relatively straightforward, basically a drawing program, was soon incorporated.

The early computer graphics monitors were monochrome and as it was important that the various layers were identifiable, different shading patterns had to be used. Likewise, when hard copy plots were made of these patterns they were originally in a single pen colour on white paper, so shading was again used. Nowadays high-quality colour monitors and plotters are readily available and the different layers are identified with different colours, although some shading may also be used in very complex processes with many mask layers. For printing convenience, the examples described below will be illustrated with shaded black and white drawings, although normally these would be in colour.

The dimensions of the various rectangles on the different layers are obviously important, for two main reasons. Firstly, of course, they define the physical size of the electronic components being generated in the fabrication process. Secondly, some overlap must be allowed in the positioning of the patterns on top of each other, because there are finite tolerances not only in the physical alignment of one mask on top of preceding patterns, but also in processes such as etching, where there may be some under- or over-etching, and diffusion, where there may be some sideways diffusion as well as in the desired vertical direction. Therefore, associated with each fabrication process is a set of design rules that specify the minimum dimensions for the proximity and overlap of various layer patterns. An important part of the design process is the layout rule checking to ensure that none of these rules has been infringed. This is particularly important if there has been any human interaction in the layout process.

In general, the layout dimensions and design rules are not specified in absolute length units, but in a relative unit, λ. This is useful when, owing to processing improvements, the size of devices and circuits can be reduced. By specifying the designs in a relative unit, the mask data can be scaled and does not have to be completely regenerated, which it would if the data was in absolute units. The value of λ is used to describe a particular process, so designers talk about a 2 μm or 5 μm process, for example.

5.5.1 Layout examples

It is best to illustrate the above points by a series of examples: these will be a couple of basic gates in different processes and will indicate the sort of drawings that are developed and how the mask patterns relate to the physical devices.

nMOS NAND gate

The NAND gate is illustrated in Figure 5.5 which shows the circuit diagram and mask layout pattern in terms of λ. There are only four mask layers shown here for simplicity: diffusion (source and drain doping definition), implantation (to define the depletion mode load), polysilicon (to define the gate structures and align them to the source and drain regions) and contact cut (to provide an electrical connection between the diffusion and polysilicon

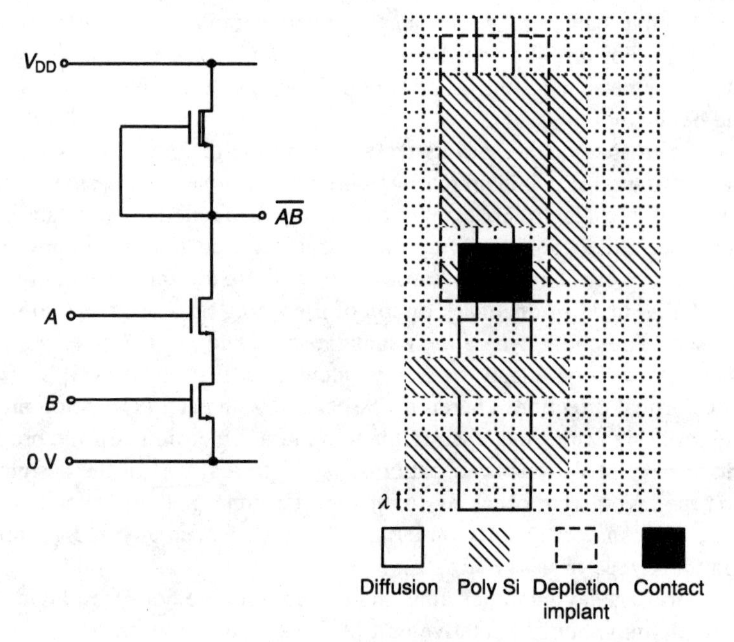

Figure 5.5 nMOS NAND gate and scaled layout

regions). The other mask layers such as metallization, passivation and other contacts have been omitted. Referring to the process description in Chapter 2, it can be seen that a transistor is formed wherever the diffusion and polysilicon patterns overlap; if this is also overlapped by an implantation region the device will be depletion mode, otherwise it will be enhancement mode. (Refer also to Appendix 1 and Figure A1.11 for a detailed description of the nMOS process.) An important design parameter, as was seen in Chapter 3, is the width/length ratio of the various transistors. As it is the ratio that is important, there is no problem in using a relative dimension, as the ratio will remain the same if the process is scaled to a smaller dimension. There are other physical effects when a design is scaled, for example the current density in the polysilicon and metal conductors, which the designer may have to take account of.

The layout in Figure 5.5 has been drawn according to layout rules relevant to the nMOS process. There are many tens of these in total, but as an illustration, just a few are now listed: minimum diffusion and polysilicon dimension 2λ; minimum spacing between two diffusion lines 3λ; minimum spacing between two polysilicon lines 3λ; minimum spacing between a diffusion and polysilicon line 1λ; minimum contact cut dimension 2λ.

This layout is effectively a scale drawing of the device as it appears on the surface of the IC. The information contained in this layout can often be of interest to a designer, although the form of the drawing is somewhat large and cumbersome. A 'shorthand' form of this layout is sometimes used which describes the topology of the layout without being a scale drawing. These are referred to as stick drawings and the equivalent of this layout is illustrated in Figure 5.6. Here all the layout rules are taken for granted, the only size information being that of the width/length ratio for the transistors which is provided in text form by each device. These stick drawings are clearly a more compact form of the layout and have the added advantage that they can be used in a mixed drawing system, involving layout, transistor and gate level schematics, as illustrated in Figure 5.7. Here the NAND gate is illustrated by

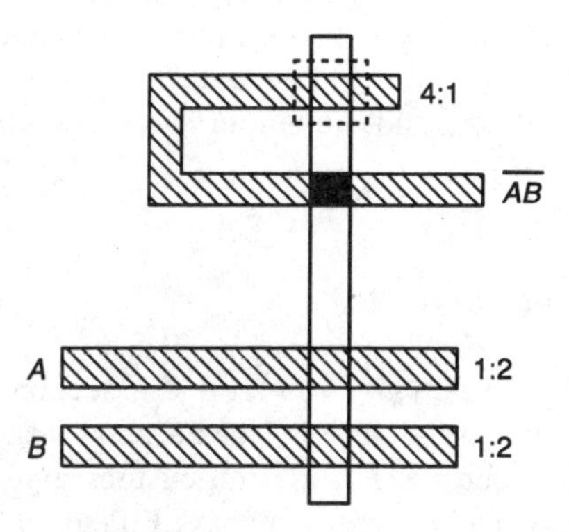

Figure 5.6 Stick diagram of NAND gate of Figure 5.5

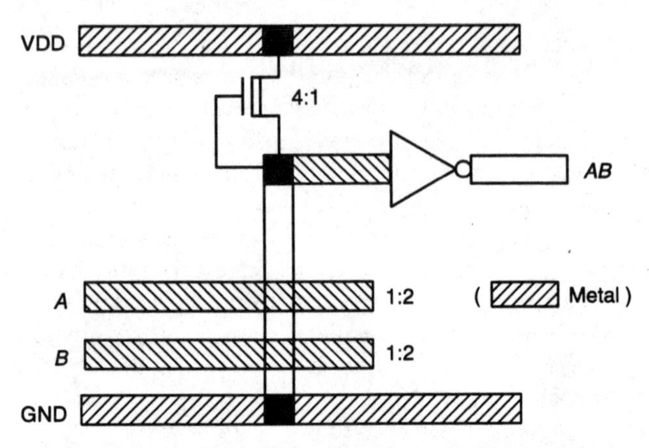

Figure 5.7 Mixed-symbol diagram of AND function

the stick drawing for the enhancement mode devices, a transistor for the depletion mode device, and it is shown driving an inverter at the gate level. Note here that the power and ground rails are also included as stick representations.

CMOS inverter

The layout of the CMOS inverter gate is shown in Figure 5.8. (See Appendix 1 and Figure A1.12 for details of the CMOS process.) This device consists of two transistors, one n-channel and one p-channel. The process illustrated here is based on a p-type substrate, so there is a requirement for an n-well doping region in which to form the p-channel device. Both transistors are enhancement mode, so there is no implantation layer needed to form a depletion mode device. The width/length ratios are different for the two transistors in order to achieve a mid-rail logic transition. This is due to the current in each transistor being associated with a different carrier type (electrons for n-channel, holes for p-channel) each having different mobilities. The ratio factor is the same as the mobility factor, ≈ 2.5, as shown by the different gate sizes. The input line is brought in on polysilicon, and the output is metal, in order to connect the different diffusion regions.

5.5.2 Automatic layout and routing

The descriptions and examples given so far in this section are most relevant to full-custom designs, where the designer has a great deal of interaction with and control over the layout process. In semi-custom approaches such as standard cell and gate array, the designer may have little or no interaction with the layout process. In the standard cell, for example, each cell has its particular

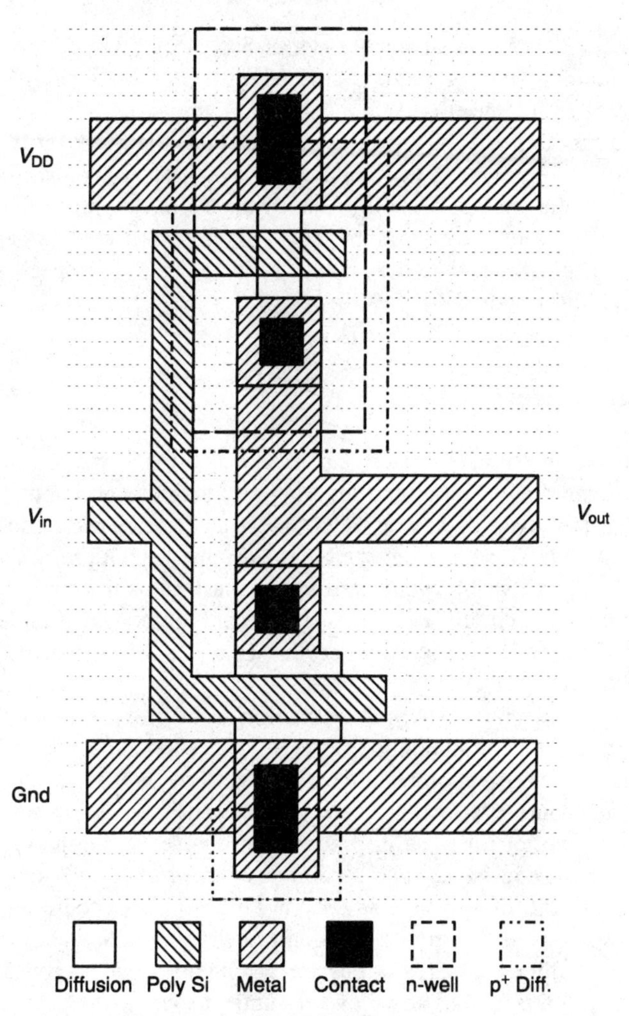

Figure 5.8 CMOS inverter layout

mask layout completely specified, the only task being to place the cells and connect up the metal and polysilicon interconnections as required. While this could be done manually by the designer, most computer packages provide an automatic placement and routing program that does the task. The optimization routine is usually based on locating the cells in such a position so as to minimize the total length of interconnections to complete the design. This is because the delay and loading associated with the interconnections can have a significant effect on the final circuit performance (see section 5.6). The designer may have some interaction in specifying cell location or altering the basis of the optimization routine. For example, certain nets may be more critical than others, and have a higher priority for the shortest length than for others. In particular time-critical signal lines such as clock lines, or nets that have a high fan-out and would therefore suffer most from extra loading, would be prime candidates for selection.

Gate array routing is a very similar process. Here the mask set is complete except for one or more metal interconnection layers. The only placement operation concerns the choice of the location of the available gates. This is again done on the basis of the minimum length interconnections. Once the gate 'placement' is achieved, the interconnection is automatically specified, based on the netlist information.

5.6 Circuit extraction

Circuit extraction (sometimes termed back annotation) from the layout patterns back to an equivalent circuit level representation is an important operation that has two main objectives. The first is an error checking operation to ensure that no mistakes have occurred in the original layout process and so the physical circuit that will be generated by the mask set is the same circuit that was designed at the circuit level. This is especially important if there was any human intervention in the layout procedure, as this is the most usual source of errors.

The second objective of circuit extraction is to provide a more accurate model of the circuit for simulation purposes. The first simulations for circuit function and test verification are almost always done on the basis that the nets joining components are ideal, of zero length, and have zero resistance connections. This is not the case in practice, of course, as the polysilicon and metal layers have a resistance and capacitance that provide additional loadings to the output of the circuit stage and, if very high speed operation is required, these may also have to be modelled using transmission line theory. The quantitative effects of these considerations are dependent on the length of the interconnection, and this is not known exactly until the whole circuit is laid out at the mask level. Once the circuit extraction has been done, the design can go through a post-layout simulation stage to check that the additional circuit effects due to layout have not affected the circuit performance to an unacceptable degree.

The most simple model of the interconnection effects is to treat them as an additional capacitive load, based on known figures for the capacitance per unit length on the particular layer and the length of the interconnect. More complex modelling can include the resistance effects of the line and examination of cross-talk effects between lines in close proximity. In very high-frequency operation, where the length of the interconnections is comparable with the signal wavelength, transmission line theory must be applied, modelling the interconnections as lines with a characteristic impedance and electrical length, and taking account of reflections due to mismatches between the characteristic and source and load impedances.

5.7 Synthesis approaches – silicon compilers

The various processes and tools involved in the creation of an IC fall broadly into two types – analysis and synthesis. Synthesis processes are associated with design creation, while analysis processes are associated with design validation. The majority of the tools so far described are of the latter type; these include simulation, HDLs, design rule checking and circuit extraction. The operations are performed on existing designs and there is no aspect of design generation associated with them. The synthesis processes we have covered are those of circuit drafting and layout.

The ultimate in computer usage in IC creation would be for the engineer to specify the complete function of the IC, in the form of a high-level functional description, and for the computer program to generate the complete mask set automatically without further human intervention. Such a process is termed silicon compilation and the program called a silicon compiler. Such tools do exist, but they are often somewhat restrictive and rarely commercially competitive compared with more interactive design methods.

The silicon compilation process is inherently a top-down design approach, as the starting point is the functional description of the circuit. Any particular silicon compiler will be based on a particular circuit architecture, since the programs are not yet sophisticated enough to choose a different type of architecture based on the original specification. A common architecture used is PLA-based, because of the ease of automatic programming once the function is in a logical or Boolean form. This is a relatively simple approach and very limited in application. It may form part of a more complex silicon compiler routine that uses different blocks of circuitry, such as PLA, shift registers, encoding circuits, etc. Each of these blocks is independently programmable or reconfigurable and the program puts together a complete circuit comprising these different programmed blocks. In this approach the automated design is therefore done at the RTL of abstraction, so the compiler must have routines for converting the chip-level functional description to the lower RTL.

Another approach to silicon compilation is to base the automated design on a standard-cell type of approach. Here the design translation has to go down yet another level of hierarchy, effectively to the gate level. This requires yet more sophistication in the compiler design routines but should lead to a more efficient design than the RTL approach. The problem with the RTL approach is that it uses relatively large blocks of programmable or reconfigurable circuit, which may be wasteful of silicon area for an arbitrary design. The standard-cell approach should lead to a much more compact realization of the circuit.

Being a fully automated design, whichever level design approach is used, the compiler will make use of place and autorouting algorithms to lay out the IC and so derive the mask data. The creation of the individual cells is likely to have been done by human designers to optimize these components, which are then library elements called by the silicon compiler.

The few existing silicon compilers are all for digital designs, since the algorithms required by analogue compilers are very complex. Research work has been done on analogue silicon compilers, but they have yet to be proved competitive compared with the custom approaches for analogue IC design.

Reference

Huang, J.S. and Taylor, G.W. (1975) Modeling of an ion-implanted silicon gate depletion mode IGFET, *IEEE Transactions*, Vol. ED-22, November, pp. 995–1000.

Bibliography

L.E.M. Brackenbury, *Design of VLSI Systems – A Practical Introduction*, Macmillan, Basingstoke, 1987.

T.E. Dillinger, *VLSI Engineering*, Prentice-Hall, Englewood Cliffs, New Jersey, 1988.

R.L. Geiger, P.E. Allen and N. R. Strader, *VLSI Design Techniques for Analog and Digital Circuits*, McGraw-Hill, New York, 1990.

P.J. Hicks, *Semi-Custom IC Design and VLSI*, Peter Peregrinus, London, 1983.

R.H. Katz, *Contemporary Logic Design,* Benjamin Cummings, Redwood City, California, 1994.

C.A. Mead and L. Conway, *Introduction to VLSI Systems*, Addison-Wesley, Reading, Massachusetts, 1980.

Meta-Software, *HSPICE Version H92 User's Manual*, Campbell, California, 1992.

G. Russell, D.J. Kinniment, E.G. Chester and M.R. McLauchlan, *CAD for VLSI,* Van Nostrand Reinhold (UK), Wokingham, 1985.

M.J.S. Smith, *Application Specific Integrated Circuits*, Addison-Wesley, Reading, Massachusetts, 1995.

W. Wolf, *Modern VLSI Design – A CAD Based Approach*, Prentice-Hall International, Englewood Cliffs, New Jersey, 1993.

Questions

5.1. Draw the circuit diagram for the transistor circuit specified in the following SPICE file. Annotate the transistors with their *W/L* sizes.

```
* FILE: PROBLEM1.SP
CMOS CIRCUIT
.OPTIONS SCALE=1E-6 WL
VIN 7 0 0 d.c. 1
```

```
.DC OP
.PRINT d.c. V(5)
M1 3 1 0 0 MN 60 10
M2 3 4 5 0 MP 100 10
M3 4 3 0 0 MN 40 10
M4 2 1 0 0 MN 60 10
M5 2 2 6 6 MP 40 10
M6 4 2 6 6 MP 40 10
VDD 6 0 5
.MODEL MN NMOS LEVEL=5 VT=1 UB=700 FRC=0.05 DNB=1.6E16
+ XJ=1.2 LATD=0.7 PHI=1.2 TOX=800
.MODEL MP PMOS LEVEL=5 VT=-1 UB=245 FRC=0.25 TOX=800
+ DNB=1.3E15 XJ=1.2 LATD=0.9 PHI=0.5
.END
```

5.2. The following is a part declaration from a SOLO 1400 HDL. Draw the circuit and label the various gates and signal lines. Work out what function the part performs.

```
Part problem2 [ina, inb]→gt, lt, eq

  Signal inva, invb
  nor [gt, lt]→eq : s1c3
  and [ina, invb]→gt : s1d2
  and [inb, inva]→lt : s1b2
  not [ina]→inva : s1b1
  not [inb]→invb : s1d1
  End
```

5.3. Write a VHDL source file for a 1-bit full adder circuit including the behavioural models of the basic gates and a structural model for the complete circuit. Use 10 ns delays for all the gate models.

5.4. Based on the layout examples of section 5.5.1 and the design in question 3.3 of Chapter 3, draw the layout of a 3-input nMOS NOR gate with $\lambda = 3\ \mu m$.

6 Testing
How can we check it works?

6.1 Introduction

Testing is an essential part of the overall realization of an IC. This is particularly so if the device is one that will be sold, either as a discrete component or as part of a larger system. For a manufacturer to ensure product quality, he or she must have the confidence that it will work correctly and reliably. There is of course a cost associated with this quality assurance, both in the time and effort involved in the testing procedure and in the extra area of the circuitry purely associated with testing functions (extra test points and any built-in self-test circuits). This latter is usually referred to as circuit 'overhead'.

In the past the subject of testing has been regarded by many engineers as a necessary evil. The circuit was designed by design engineers and then passed on to test engineers to devise suitable test routines. This was often the first time that the test engineer was involved with the product. As circuit size and complexity grew, this manufacturing route was found to be too inefficient and there was a realization that testing was an integral part of the overall design and that test engineers should be involved from the start. Indeed many engineers developed a dual role of design and test, and the concepts of 'design for testability' (DFT) evolved. Today even the term 'overhead' is regarded as negative, as it indicates that the parts of the circuit dedicated to test play no useful function. Instead engineers are being encouraged to view these elements in terms of quality assurance and hence more positively (while still attempting to keep these aspects to a minimum!).

The types of tests that can be performed on any particular circuit are legion. This is probably more so in the case of analogue circuits where the variables are continuous in nature. The sort of test to be applied to a circuit clearly depends on what function the engineer wants the test to perform. Very broadly speaking these can be categorized into three areas:

Development test. This aspect is for the verification of a new design and/or process. The main aim is to check that the design, which may well have been verified theoretically using a suitable simulator, can be successfully translated into a working circuit. The design may be a new one, or possibly an existing one that has been translated into a different process (either based on a different technology or a different minimum feature size, λ). In any event, it is vital that the prototype operation is thoroughly tested before being moved into production, so as much time and expense as are necessary should be devoted to this stage of testing. At this stage, the engineer is testing for fundamental design or process faults that would prevent the circuit from functioning properly over its entire design specification range. If such a fault is not detected until the production stage, the cost in wasted products can be immense. As well as functional test, some diagnostic test may be involved in order to determine the source of any non-functionality.

Production test. Once the design has passed through the prototype stage and is being produced in large quantities, the emphasis of the testing operation is changed. Now the engineer is no longer looking for design or fundamental process faults but for localized fabrication faults that will cause failure of the circuit. The financial balances of IC production were outlined in Chapter 1, the important aspect here is that testing accounts for a certain percentage of the overall product cost. It is uneconomical to go into the depth of testing associated with prototype testing – what is required is a relatively short test sequence that results in a simple go/no-go decision. The optimization of the testing procedure to minimize the test effort while ensuring reliable detection of faulty circuits is the heart of the problem, and much work needs to be devoted to meet these objectives.

Field test. Many of the IC failure mechanisms are dependent on variables such as temperature, moisture and electrostatic discharge, and so a circuit that may have been functioning correctly when first manufactured will fail at some later time. If the circuit is part of some larger system then it is not feasible to scrap the complete system because of the failure of one small part. However it is often not obvious which is the failed part, and this has to be determined by some field or diagnostic test. Large systems are constructed of smaller subsystems, often mounted on easily removable boards. This enables the engineer to locate rapidly the faulty board, if not the individual circuit, and replace the complete board to get the system up and running as quickly as possible. The fate of the board will depend on the particular economics of the sub-circuit and the ICs on it. Often the whole board will be scrapped, as the cost of the diagnostic test to locate the faulty component and the subsequent re-work is higher than that for simply manufacturing a new board. If the economic equation works in the other direction, the board may well be re-worked. Some help in this diagnostic testing can be designed into the ICs by having built-in self-test (BIST), in which the IC can be switched to a self-testing mode to check out the functional part of the circuit.

Perhaps the most critical of these three types of testing, and certainly the one to which most effort has been applied, is that of production test. This is the aspect on which the rest of this chapter concentrates.

6.2 Fault modelling and test strategies

Before a strategy for test can be constructed, the engineer has to be aware of what the test is meant to determine. In the case of production test this is to check for the presence of fabrication faults that would cause a functional failure in the resulting electronic circuit. As can be seen from the number of different processes involved in fabrication (refer to Appendix 1), the number of potential faults is enormous, although the efforts made in process control should mean that in reality very few faults actually occur. Indeed the majority of circuits are fault free. The ratio of the number of faulty circuits in a processed batch to the total number of circuits fabricated is termed the yield. Yield is a function of the chip area: the larger the IC the lower the yield and, given a particular circuit and process, there is a minimum yield below which the fabrication becomes uneconomical. Another assumption often made in the design of test strategies is that a maximum of one fault occurs in any given circuit.

6.2.1 Fault modelling

Typical processing faults that can occur are holes in insulating layers, bridging connections between tracks (usually in metal layers), missed contacts, poor control of doping or etching, electrostatic discharge breaking down dielectric layers, and several other mechanisms. There is a critical step in relating the physical processing fault, whatever it might be, to the way it affects the electrical performance of the circuit component being formed. This will depend both on the fault and the particular component. The translation of the processing fault into an equivalent electrical circuit for the resulting faulty component is termed fault modelling.

Broadly, faults can be categorized into two distinct groups. The first is catastrophic or hard faults. As the name suggests, these cause a complete failure of a particular component, and so include such effects as short and open circuits. The second class are parametric or soft faults. Here the device may still operate in its normal way, but the parameter values may be altered to such a degree that the circuit as a whole does not function correctly. For example, the values of resistance, capacitance or transistor gain may be changed beyond their normal tolerance values.

6.2.2 Testing strategies

As was mentioned above, the main purpose of production testing is to verify the correct operation of a circuit, given the possible presence of certain processing faults. In other words, we are checking that the circuit functions correctly and this testing approach is termed 'functional testing'. In order to be certain that the circuit is completely functional, it should strictly be tested under all possible operating conditions, but this is where we run into a problem with large digital circuits. All possible operating conditions in this case include all the possible states that the circuit can be in. Even for large combinational logic circuits the number of possible states can be huge, but if the circuit also contains memory elements then the task becomes impossible. For example, if a finite state machine has n primary inputs and m lines fed back from output to input, the total number of possible states is 2^{n+m}. Say $n = 24$ and $m = 20$, then this results in 2^{44} possible states. Testing at a rate of 10^6 per second means it would take 6 months of continuous test to test the circuit exhaustively. This situation can be helped by partitioning the circuit into smaller blocks and performing separate tests on each block, but this is at the expense of providing access to internal nodes in the circuit by way of extra probe points.

An alternative approach to this problem is to consider the test strategy from a different viewpoint. Here the assumption is made that the design of the circuit is correct and with no processing faults the circuit will function correctly. So instead of testing for functionality, we systematically test for the presence of processing faults. Such an approach is termed 'structured testing'. The problem is now put into the area of fault modelling, as we have to test for the faults from an electronic-circuit point of view.

6.2.3 'Stuck-at' faults

It has been found that, in the case of digital circuits, the majority of process faults can be modelled at the circuit level by the concept of 'stuck-at' faults. Here the output from a gate is considered to be stuck at either logic 1 or logic 0, irrespective of the state of the inputs to the gate. So the faults we are testing for are nets being 'stuck-at-1' ('s-a-1') or 'stuck-at-0' ('s-a-0'). For example, if the output line of a gate has a short-circuit fault that connects it to the signal ground, then that net will be stuck at logic 0 under all conditions.

The number of faults to be tested for now depends on the number of nets in the circuit, and test signals (test vectors) have to be designed to test for each of these nets being s-a-1 and s-a-0. In practice the total number of faults to be tested for is less than twice the number of nets, because certain stuck-at faults on different nets cause the same effect. This number is usually many orders of magnitude less than the number of functional states that the circuit can be in.

An example of how a gate may be tested for the presence of a stuck-at fault, and how different faults are identical, is illustrated in Figure 6.1. Figure 6.1a shows the test for the output of an AND gate s-a-0. If 1s are applied to the two inputs, the result should be 1, but if the output is s-a-0 we will measure the erroneous result of 0. However, Figure 6.1b shows that this is an identical test and result for one of the input lines being s-a-0. Here we are trying to apply two 1s to the inputs, expecting a 1 at the output, but as one of the inputs is 0, the output will again be the erroneous 0.

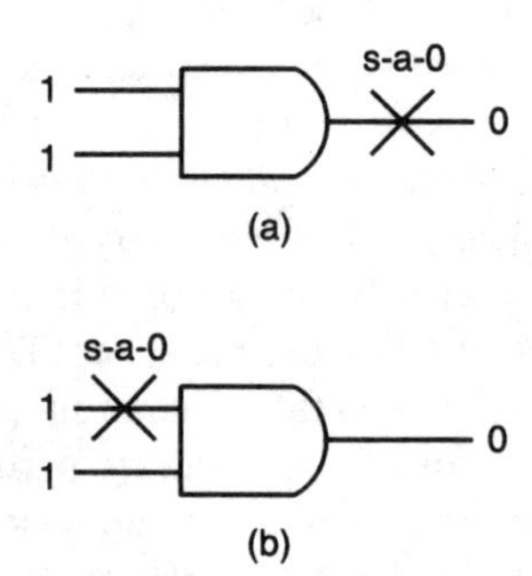

Figure 6.1 Stuck-at fault testing: (a) test for stuck-at-0 output of an AND gate; (b) demonstration of two stuck-at faults being identical

6.2.4 Controllability and observability

The example of Figure 6.1 leads neatly on to the key concepts of testing and testability – those of controllability and observability. The tests illustrated in the figure are fine when considered in isolation, however in a real circuit the gate may well be deeply embedded among other circuitry and we cannot directly access the nets at the inputs and outputs. Often there is only access to the primary inputs in control and the primary outputs to observe. In order to set the test conditions up, we must be able to set the primary inputs into conditions such that the appropriate signals occur at the inputs to the gate. Likewise, we must be able to transmit the result of the test to an observable output. If either or both of these conditions are not possible, then the fault under consideration is not testable. The ability to be able to achieve the former condition is termed 'controllability' and the ability to achieve the latter condition is termed 'observability'.

The number of testable faults as a percentage of the total number of possible faults is termed the fault coverage. This factor depends on the signals used at the circuit inputs (the test vectors). A critical consideration here is that of speed. As testing is a relatively expensive process, it is best to make the test as short as possible. However, the shorter the test sequence, the less the number of test vectors and the lower the fault coverage, so there is a trade off between length of test sequence and fault coverage, and much work has been done to develop test pattern generation routines in order to optimize the trade-off. Typically, fault coverage of the order of 95 per cent is acceptable for a large digital circuit, while testing times should be of the order of one or two seconds.

6.2.5 Test pattern generation techniques

Based on the single stuck-at fault condition, it is possible to determine the required test pattern to determine any particular fault, provided that the fault is testable. Two such methods are described below. This process can be repeated for each possible fault, resulting in a maximal set of test vectors. As has been seen, any specific test may cover several faults, so it may be possible to reduce the set of test vectors and still maintain total fault coverage. The process of test vector reduction, sometimes referred to as 'fault collapsing' is a pre-process stage implemented before test pattern generation. Typically, the fault collapsing process will halve the number of test vectors required. It may still be prohibitively long to implement this collapsed set and a rigorous selection of the reduced test set is computationally very long; so minimal test vector sets are chosen based on the compromise of minimizing test time against maximizing fault coverage.

A description now follows of two test vector generating algorithms: D-algorithm and Boolean difference.

6.2.6 D-algorithm

The D-algorithm is due to Roth *et al.* (1967). It is valid for non-redundant combinational logic circuits only, although the algorithm may be modified for sequential circuits. The algorithm is based on a structured approach to the test vector search and comprises four steps:

(1) Fault excitation. Inputs are set up so that the net under consideration is driven to the logical value which is the opposite to the effect of the fault. Figure 6.2 shows an example in which the test for net E being s-a-0 is to be generated. We require E to be driven to logic 1, which requires A and B both to be set to 1.
(2) Fault effect propagation. To make the result of the test observable, the effect has to be propagated to a primary output. In the example, the result has to be propagated to the output Z which requires G to be set to 0 and D to be set to 1 to ensure the outputs of the OR and AND gates follow the input lines with the fault effect. This conditioning of inputs to create a path

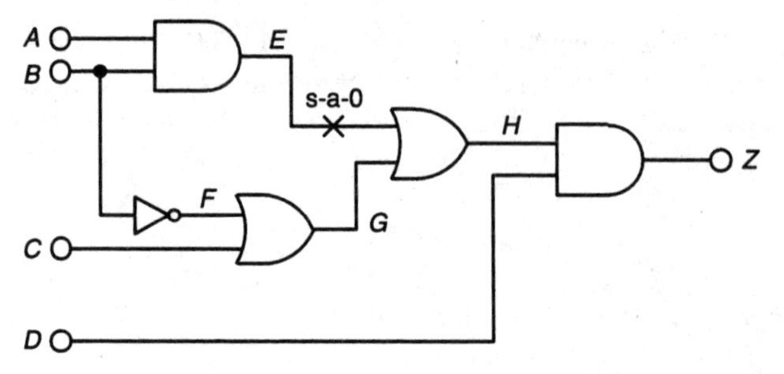

Figure 6.2 D-algorithm example

through the circuit along which the test result is propagated is termed 'path sensitization'.

(3) Line value justification. The implication of the path sensitization on the primary inputs must be examined by backtracking in order to check for any contradiction in the conditions of step 1. The condition $D = 1$ causes no problem, as D was not set in step 1. The other condition from step 2 is $G = 0$. This implies that C and F are both set to 0 and therefore B is set to 1. This is consistent with the conditions in step 1 where B was then set to 1. The values are justified and the test vector is $A = 1$, $B = 1$, $C = 0$, $D = 1$. Measurement of Z gives the result that if $Z = 1$: fault not present; if $Z = 0$: fault present. If there was an inconsistency in the primary inputs from steps 1 and 2, step 1 would have to be repeated with an alternative input specification to excite the fault. If no alternative exists, the fault is deemed undetectable. For example, if the inverter were not present in the circuit of Figure 6.2 (B connected directly to F), this fault would be undetectable.

(4) Line value consistency. Once the primary input values have been determined, it is necessary to track forward through the circuit to check for any problems with multiple path sensitization and reconvergence which can mask the propagation of a test result to an observable output.

The problem with the D-algorithm is that in general, for larger circuits than this example, there are many possible paths that can be sensitized and the chosen path in step 2 is arbitrary. The effort required in back and forward tracking and possible repeats of steps 1 and 2 can be significant. However, if a test exists, the D-algorithm is guaranteed to find it eventually. Some refinements can be made to the algorithm in terms of pre-processing of the circuit structure and responses to reduce the backtracking effort, but the choice of sensitized path is still arbitrary.

6.2.7 Boolean difference

The Boolean difference algorithm is an analytical approach that generates all possible tests for a given fault. It is based on a form of differential calculus which in Boolean algebra is achieved through the exclusive-OR operation.

Consider a section of combinational logic which has n primary inputs. The output function Z can be written in the form

$$Z = \mathcal{F}(X_1,\ X_2,\ \ldots,\ X_k,\ \ldots,\ X_n)$$

where X_i are the primary inputs. Now suppose a stuck-at fault is present in the input X_k. The output function now becomes

$$Z_k = \mathcal{G}(X_1,\ X_2,\ \ldots,\ \overline{X}_k,\ \ldots,\ X_n)$$

which is formed by replacing X_k with $\overline{X}_k$. The Boolean difference function is defined as

$$\partial Z/\partial Z_k = Z \oplus Z_k = \mathcal{H}(X_1,\ X_2,\ \ldots,\ X_n)$$

where $\oplus$ is the EX-OR function. The result of the Boolean difference calculation results in the complete set of tests for the stuck-at faults. The implementation will be demonstrated by means of an example. Consider the circuit shown in Figure 6.3. The fault-free Boolean function is $Z = X_1X_2 + \overline{X}_2X_3$. If line A has a stuck-at fault, then we replace the variable X_1 with $\overline{X}_1$ to derive $Z_{X_1} = \overline{X}_1X_2 + \overline{X}_2X_3$. The resulting Boolean difference function is given by

$$\begin{aligned}\partial Z/\partial Z_{X_1} &= Z \oplus Z_{X_1} \\ &= (X_1X_2 + \overline{X}_2X_3) \oplus (\overline{X}_1X_2 + \overline{X}_2X_3) \\ &= X_1X_2X_3 + X_1X_2\overline{X}_3 + \overline{X}_1X_2\overline{X}_3 + \overline{X}_1X_2X_3\end{aligned}$$

The four minterms derived above define the full set of input tests that will detect both types of stuck-at faults. To determine which tests determine which fault, the tests are separated into those tests containing X_k and those containing $\overline{X}_k$. The former will require a 1 on X_k and so detect s-a-0 and the latter require a 0 on X_k and so detect s-a-1. In the above example, $X_1X_2X_3$ and $X_1X_2\overline{X}_3$ detect s-a-0 and $\overline{X}_1X_2\overline{X}_3$ and $\overline{X}_1X_2X_3$ detect s-a-1.

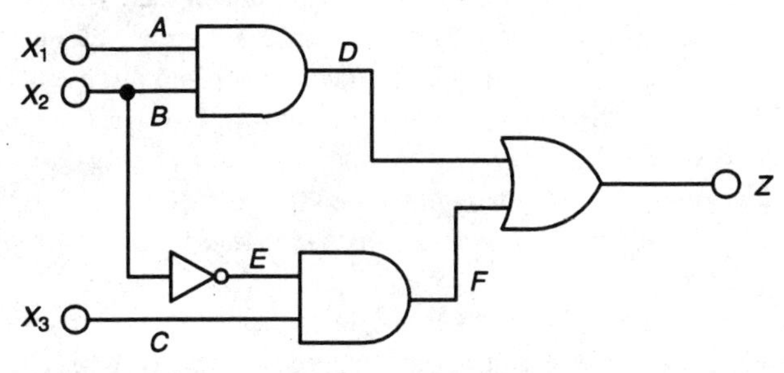

Figure 6.3 Boolean difference example

The routine illustrated above is only for faults on primary inputs, but it is an easy matter to extend the technique to faults on internal nodes. In this case the output function Z has an extra variable f_k which is dependent on the primary inputs and represents the Boolean function of the internal node under consideration. Taking the Boolean difference of Z will result in minterms involving f_k which represent the required tests.

The Boolean difference technique can be extended to multiple as well as single fault detection. It can become rather slow and unwieldy for large circuits where the EX-OR function of very complex functions has to be performed, but it is often used to determine the faults in small parts of circuitry and can then be used in conjunction with other techniques, such as the D-algorithm, when applied to large circuits.

6.2.8 Other fault models

All the preceding techniques have been based on the stuck-at fault model. This model has been adopted as the most common, and most commercial automatic test pattern generation (ATPG) software is based on it. However, there are several other physical faults that can occur which are not well modelled by the stuck-at approach. The shortcomings of the stuck-at fault have long been recognized and, although some research work has been done at devising tests for other faults, they have rarely been incorporated in ATPG. This is because the nature of the faults make them difficult to test or because the proliferation of possible faults makes the test uneconomically long to perform. A brief description of some of these other faults follows.

Stuck-open fault

The stuck-open fault, like the stuck-on fault below, is characteristic of CMOS circuits. It occurs when a fault in an MOS transistor causes it to be permanently in the high-resistance state, irrespective of the input state. In simple nMOS circuitry this fault can be successfully represented by the single stuck-at model. However in CMOS the fault can cause a memory element to be created, turning the circuit from a combinational circuit into a sequential one. The effect is illustrated in the example shown in Figure 6.4. Here we have a two-input NAND gate in CMOS, the stuck-open condition being in transistor P_2. The capacitors represent the loads presented to the gate by the inputs of other CMOS gates. Although usually small in value, these capacitors play a vital role in stuck-open faults.

If the gate is to be fully tested for single stuck-at faults, the set of test vectors to be applied, and the fault-free responses, are given in the following table:

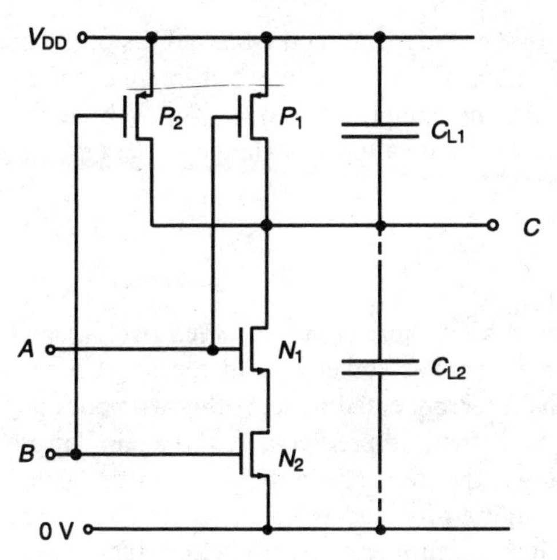

Figure 6.4 Two-input CMOS NAND gate

A	B	C
1	1	0
0	1	1
1	0	1

However, this sequence will not detect P_2 stuck-open, which should normally be detected by the third test vector. However, when the second test vector is applied, the 1 at C is retained by the charged capacitor so the apparently correct output response is observed. In order to test for this fault, the order of testing is important, and to check completely for stuck-open faults the following test vectors must be applied in the correct order, resetting the output at each stage so that no signals are retained by the capacitors:

A	B	C
1	1	0
0	1	1
1	1	0
1	0	1

Effectively the test for stuck-open faults is two vector – a reset vector followed by a test vector on the next clock cycle.

Stuck-on fault

A similar fault to stuck-open faults is the stuck-on fault where a transistor is permanently switched on. The effect is more subtle and difficult to test for, as the gate response will depend on the actual transistor dimensions. This is because, if one transistor is permanently in a low-resistance condition, the voltage at the output will depend on the resistances of the other devices forming a potential divider. The voltage may fall in either the logic 1 range or the logic 0 range. It is almost impossible to test for such faults by the normal test vector approach unless a detailed knowledge of the process is known.

Bridging fault

This fault is caused by a short circuit between two adjacent lines, usually in the metallization layer, but sometimes in the polysilicon layer. The actual logic value on the interconnected node, if the two nodes are driven by different values, will be difficult to predict as it will again depend on the particular devices' technology, the transistor dimensions in the gates involved, as well as the line and bridging resistances. This again makes testing with normal test vectors very difficult. A further difficulty with bridging faults is that the location of the short, in net terms, is dependent on the circuit layout. Nets at opposite sides of the IC are highly unlikely to be bridged. The layout information is not available at the schematic stage where test patterns are often devised. To test for all possible net bridges is wasteful as there are likely to be thousands of combinations, the vast majority of which would never occur. So if bridging faults are to be tested for, the test pattern must be devised after the layout stage, which does not fit well with the normal design procedure. A compromise is sometimes adopted, whereby bridging faults between adjacent pins only are considered.

Gate oxide short (GOS)

This fault occurs when there is a short between the gate conductor through the thin oxide layer to the transistor channel below. The behaviour of such a fault is again very difficult to predict as it depends on the position and size of the shorted area and the resistance of the shorting link.

6.3 Design for testability

As has been mentioned earlier, the key aspects of test are controllability and observability. To enhance the testability of a circuit, one must improve either or both of these factors. There are a number of techniques for enhancing test, some of which may simply be regarded as good engineering practice, but there are also specific approaches to improve testability. All these aspects fall into the broad heading of design for testability (DFT).

It has long been recognized that it is insufficient for the IC designer to concentrate purely on the functional design and to consider the test aspects as 'somebody else's problem'. To make a product economically viable, the design and test have to go hand in hand and testability considerations have to be made from the outset. This section outlines some of the techniques and tools now available to the IC engineer to make the product viable and to enhance its quality.

6.3.1 Test enhancements

The most obvious way to improve controllability and observability is to provide direct access to internal nodes. This is obviously at the cost of providing extra input and output pins. If the IC is 'core-limited' rather than 'pad-limited' then extra pads can be added with virtually no increase in chip area and hence cost. Figure 6.5 illustrates the difference between a core-limited IC and a pad-limited one. The extra chip pads can be left purely for probe test, or if there are also spare pins in the device package, these pads can be bonded in so that they can be used for field as well as production tests.

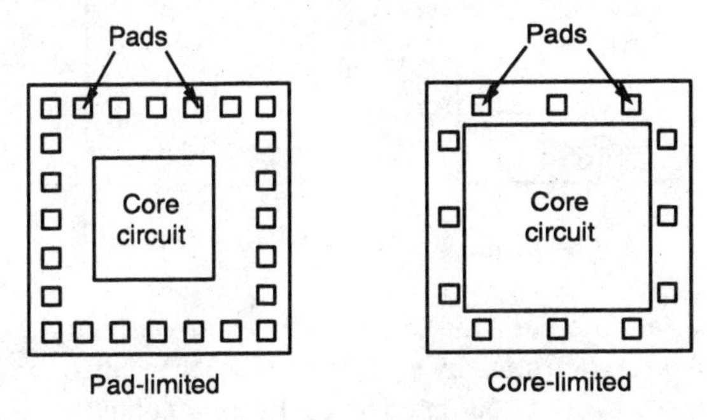

Figure 6.5 Difference between pad-limited and core-limited ICs

For purely observing the internal node, the signal is simply taken to the spare pad. If the node is to be controlled for test purposes, two pads are required: one for the test input and one to disable the normal signal to that node. One way of achieving this control is to place two NAND gates in series in the node to be controlled, as shown in Figure 6.6. When $A = 1$ and $B = 0$, the circuit function is unchanged, provided that the additional gate delay does

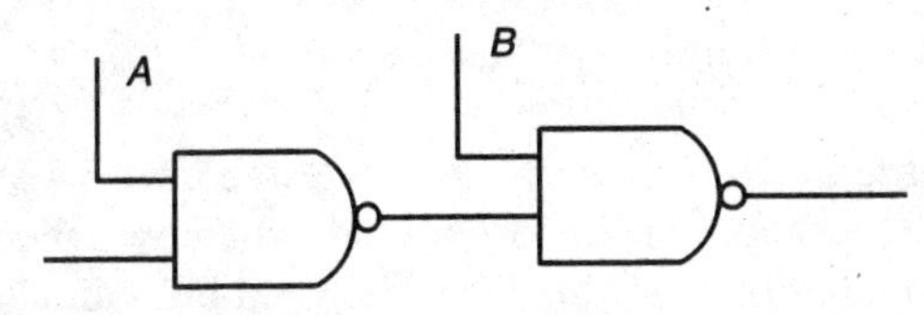

Figure 6.6 Degating a node for control purposes

not introduce problems. With $A = 0$, the normal signal is disabled and the output follows the test input B. This technique is termed 'degating' and is a form of circuit partitioning. Wholesale partitioning of ICs into different blocks and subsequent separate testing are the basis of many other DFT techniques, as will be seen later.

If the IC is pad-limited, or it is inconvenient to add extra pads or pins, an alternative approach to enhance testability is to use the same primary input and output pins for different functions. This will involve the use of extra circuitry, however, in the pad-limited case, where there is vacant chip area, the cost of the test 'overhead' is minimal. The most common approach to shared pins is through the use of multiplexers (MUXs). The principle of this technique is illustrated in Figure 6.7. The first of the multiplexers, controlled by

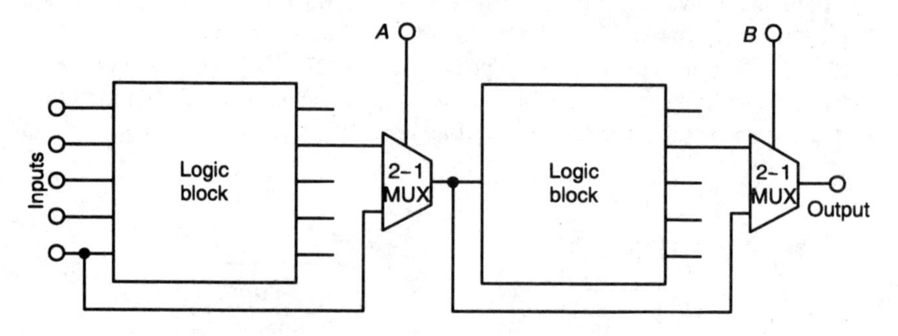

Figure 6.7 Shared use of I/O pins

signal A, allows an internal node to be driven from a primary input, while the second multiplexer, controlled by signal B, allows the state of an internal node to be observed at a primary output. The controlling conditions are as follows:

(1) $A = 0$, $B = 0$. The circuit operates in its normal mode, the output from the first logic block is passed to the second logic block, and the output from this goes to the primary output.
(2) $A = 0$, $B = 1$. The second logic block is bypassed, and the state of the internal node is passed through the second multiplexer to the primary output.
(3) $A = 1$, $B = 0$. The first logic block is bypassed, and the state of the internal node is controlled by passing the primary input through the first multiplexer.
(4) $A = 1$, $B = 1$. In this case the primary input would be passed straight to the output, so this condition is never used.

In general a large block of circuitry is less testable than a smaller block, so a 'divide-and-conquer' approach is often used, whereby a piece of circuitry is partitioned into smaller blocks which can be tested individually, as mentioned earlier. Multiplexers can be used to partition circuits into separate blocks for testing purposes, as illustrated in Figure 6.8. Here the normal interconnection of the two logic blocks is broken by the multiplexers MUX1 and MUX2. These signals are also taken to primary outputs, as they represent outputs from the separated logic blocks. Thus each of these blocks can effectively be tested in isolation.

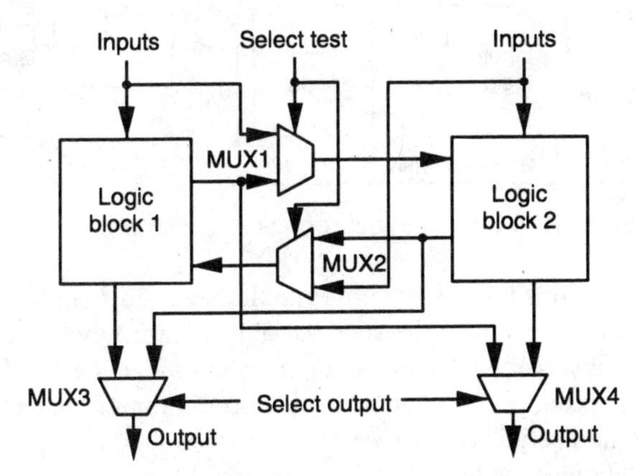

Figure 6.8 Use of multiplexers to partition combinational logic blocks

Sequential circuits, particularly with asynchronous feedback paths, are notoriously difficult to test. In this case the testing task can be eased by breaking the feedback loop. This can be done by using multiplexers or by degating techniques.

The techniques we have looked at so far in this section are only making the best of a bad job, that is they are ways to improve the testability of a circuit to which no thought of DFT was previously applied. They are sometimes referred to as *ad hoc* techniques, as the exact approach taken depends on the individual circuit under consideration. We now look at more formal, or 'structured' methods for designing in testability from the outset.

6.3.2 Scan path approaches

The enhancements to testing combinational blocks of circuitry and the automatic test pattern generation are relatively straightforward and have been described above. The major difficulties of testing arise in sequential logic structures. The problem here is that as well as the usual, easily controllable inputs to a block of logic, there are often fed-back state variables, over which very little control can be exercised. The general structure of a synchronous sequential circuit is shown in Figure 6.9, and consists of three main blocks. The input and output logic blocks are combinational in nature, but the flip–flops that derive the state variables are the complication. As the circuit stands, the state variables are neither controllable nor observable and as each of the combinational blocks has some state variable inputs, none of the blocks is properly testable.

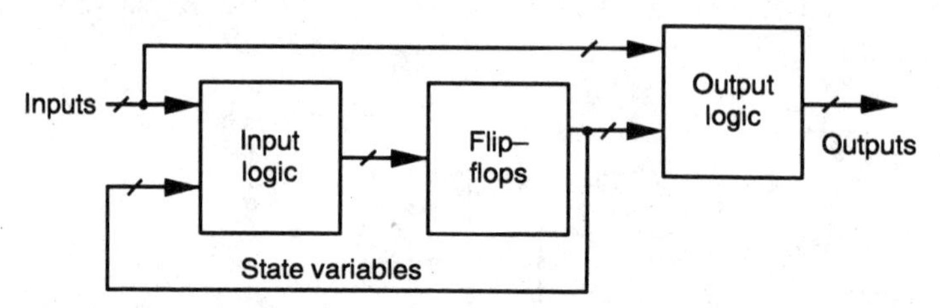

Figure 6.9 Structure of synchronous sequential logic circuits

In order to make this basic structure testable, we need to be able to isolate the combinational blocks from the state variable storage devices and so properly control and observe the signals. In order to minimize the number of additional input and output pins, the control and observation of the state variables should be done in a serial way, so we create a serial data path or 'scan path'. The implementation of the basic scan path is straightforward and again makes use of multiplexers as illustrated in Figure 6.10. Here we are just considering the block of flip–flops, which for the sake of convenience is considered to be a series of individual D-types. All the multiplexers are controlled by a single bit line. With this control bit set to 0, the output of the multiplexers are the outputs from the combinational logic block and the circuit operates in its normal functional mode. When the control line is set to 1, the output of each multiplexer is from the output of the previous flip–flop, so the previously parallel set of flip–flops is now configured as a serial shift register, generating the scan path. Data can be scanned in through the input to the first (bottom) multiplexer or scanned out from output of the final (topmost) flip–flop. This system is therefore sometimes referred to as a scan-in, scan-out (SISO) structure.

The test sequence is as follows. To control the state variables, the control line is set to 1 and the required pattern of state variables is scanned in. This will take as many clock cycles as there are flip–flops, the disadvantage of a serial data system. Once these are set, the control line is set to 0 (normal circuit operation) and the primary input sequence applied. One more clock cycle is applied to latch in the resulting outputs from the first logic block. The control line is then set to one and the results scanned out. Likewise the primary outputs from the output logic block can also be observed, so the inputs and outputs of all three circuit blocks can be controlled and observed.

There are three extra lines in this system: the control line, the scan data input (SDI) and the scan data output (SDO). The testing sequence is such that while data is being scanned in or out, the primary inputs and outputs are not being used, so it is possible to multiplex the SDI with a primary input and the

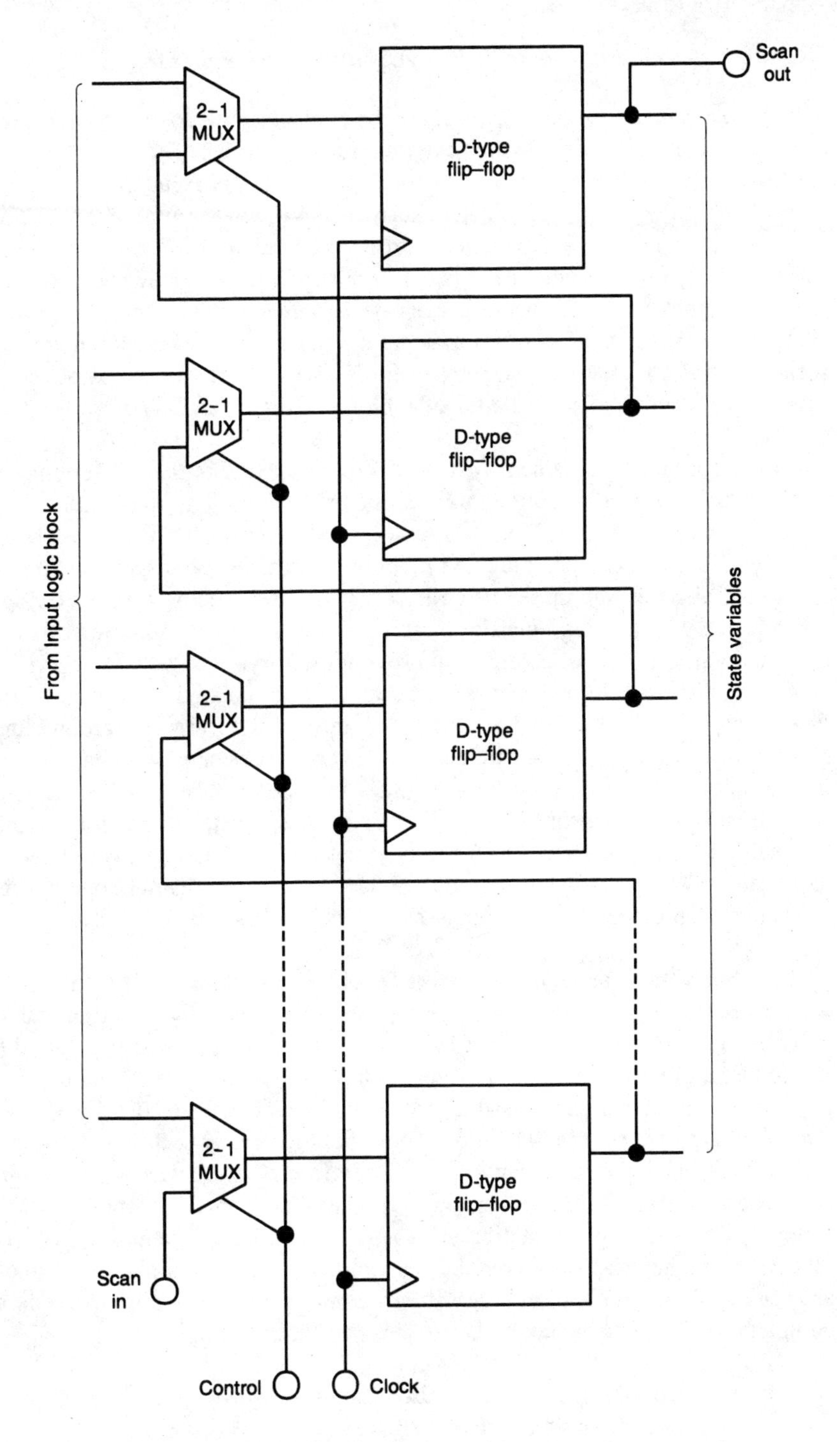

Figure 6.10 Scan path construction

SDO with a primary output, again under the control of the control signal, saving two external pins if these are at a premium.

There are a number of variations of the scan path approach. One, originated by IBM and termed level sensitive scan design (LSSD), involves the use of a shift register latch (SRL) which performs the dual function of multiplexer and flip–flop. It operates with a master–slave flip–flop configuration and two-phase clocking system to ensure correct functioning of the system, being tolerant to timing delays on the input signals. Another variation, known as scan-set, does not make use of the system flip–flops, but has a separate scan path through which internal nodes of a block of logic can be controlled and observed.

One important development of the scan path system is that of 'boundary scan'. This is primarily applied to circuit board testing. With the increasingly higher packing density of ICs on a board, together with the use of surface-mount devices and devices mounted on both sides of the board, the testing of boards has also become more difficult. The traditional probing systems cannot always cope with this high density of components, hence the adoption of scan path techniques in board testing to improve testability.

ICs to be mounted on the board have an SRL associated with each of its functional pins. These can be configured into a shift register around the boundary of the IC. There is an SDI and SDO pin associated with each IC as well as appropriate control and clock lines. Included on the PCB, as well as the normal interconnection between the ICs, is a scan path connection so that the scan registers within each IC can be connected to form an overall serial data path with a board scan input and a board scan output. Other sophisticated systems can be devised to isolate individual ICs from the board level scan path.

When boundary scan techniques were first being investigated, the principles were readily accepted by the engineering community. However an immediate problem arose in that, for a board level scan path to be successful, all of the ICs to be mounted must have the same architecture of SRL and internal scan path control. As it is highly possible that a board may consist of ICs from more than one manufacturer, the situation would be hopeless if each manufacturer used a different boundary scan system. Therefore a group was organized in 1985, the Joint Test Action Group (JTAG), to establish an acceptable standard for boundary scan implementation. This resulted in an IEEE standard, 1149.1, recently being adopted. This covers a standard test access port and the boundary scan architecture, and although it is the only ratified standard, there is a complete family of such standards in the process of development:

1149.1b Boundary Scan Description Language (BSDL) which is used in conjunction with 1149.1
1149.2 Extended serial–digital interface standard

1149.3 Real-time bus standard [work on this has recently been disbanded]
1149.4 Mixed-signal test bus standard
1149.5 Module test and maintenance bus standard

Standards from each of the working groups involved (except 1149.3) are expected to become available in the near future.

6.3.3 Built-in self-test (BIST)

While scan path techniques enhance the testability of otherwise difficult, or impossible, to test circuits, they still rely on the test vector input and observation being performed externally. Another technique of structured DFT is to include other elements of the test process directly into the IC. This may include any or all of the elements of test vector generation, control, observation and verification. Such circuitry is termed 'built-in self-test'(BIST).

There is much more scope for incorporating BIST at the board or system level; for example, test vectors could be stored in a separate ROM and called up when required, or test control algorithms or data could be stored as microcode for processor systems. The degree to which BIST is incorporated into a single IC is largely governed by the economics of the complete manufacturing process. BIST is necessarily an overhead consideration and therefore costs money in terms of chip area. To use a large proportion of the chip for test purposes, it must be shown that the cost of other required testing is more than recouped. The systems of BIST that are most often incorporated into individual ICs are PRBS test vector generation, signature analysis and built-in logic block observers (BILBO).

For test vectors to be tailored to a particular circuit and generated on chip would involve an excessive amount of overhead. Instead a very simple technique is used to generate a test vector pattern of 1s and 0s which is apparently random in nature. These patterns are termed pseudo-random binary sequences (PRBS) and are easily generated using a linear feedback shift register (LFSR). An example of this circuit is illustrated in Figure 6.11. This consists of a number of flip–flops configured as a synchronous shift register, with various flip–flop outputs 'tapped' and operated on in an exclusive-OR function. The resulting signal is then fed back to the input of the shift register, so the whole circuit becomes a signal generator. The pattern resulting from any particular configuration appears to be random, but is of course entirely predictable and is also cyclic – the PRBS will repeat after a certain length. The length of the sequence depends on the number of shift register stages, *n*, and the position and number of the taps. The maximum length of sequence before repeating is given by $2^n - 1$, that is all possible combinations except one – the all-zeros state. Using the all-zeros condition as a starting point results in a continuous pattern of zeros and is therefore to be avoided. It has been shown that

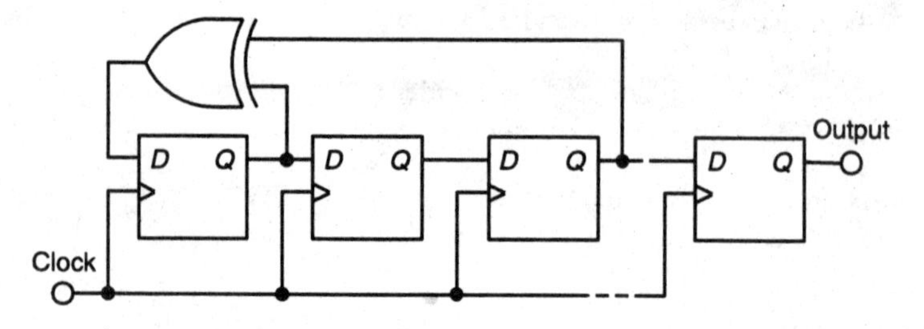

Figure 6.11 Linear feedback shift register configured as a PRBS generator

with maximal-length PRBS sequences used as test vector inputs, very good fault coverage can be achieved for a very small circuit overhead.

For a production go/no-go test, the output signals resulting from a particular test vector stimulus must be compared with the expected pattern from a known good circuit, usually derived by simulation. This comparison can be very complex and difficult to achieve on chip. However, use can again be made of LFSRs in a technique termed signature analysis. Here, the results from a monitored node are included in the LFSR, as shown in Figure 6.12.

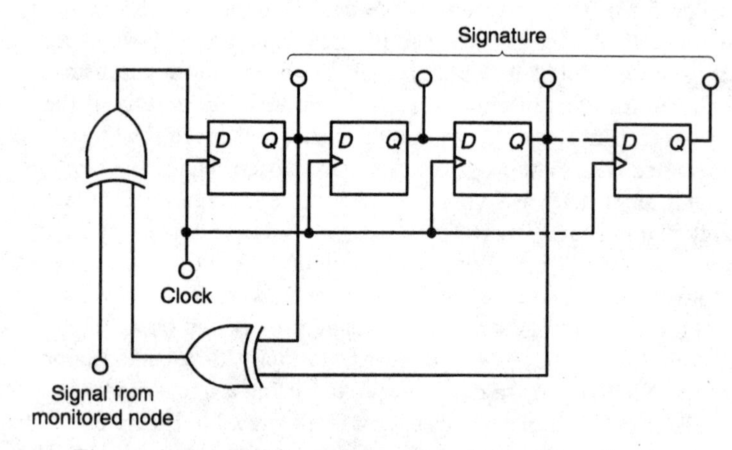

Figure 6.12 LFSR configured as signature analyser

Given the expected pattern from the monitored node, then, after the system has been run for its full sequence length, the register will contain a pattern of 1s and 0s that is characteristic of that particular node. This is termed the signature. Checking that the correct signature is present is then a very straightforward operation. Provided that the sequence length is sufficiently long, the probability of an incorrect output pattern giving rise to the correct signature becomes vanishingly small. The system can also be adapted to analyse several output nodes simultaneously, as illustrated in Figure 6.13. Here each of the monitored nodes feeds into a different stage of the register.

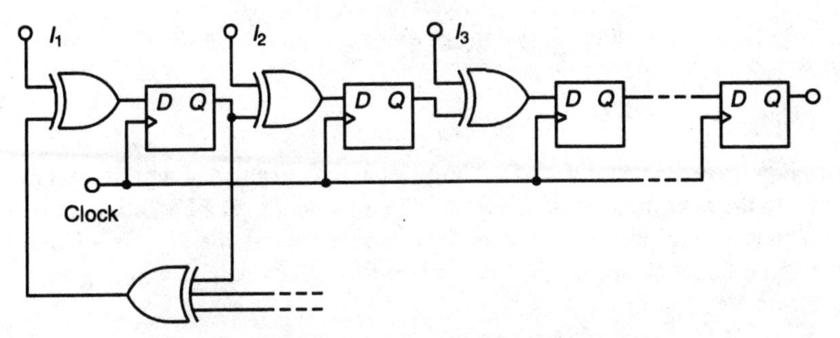

Figure 6.13 Multiple input signature analyser

BILBO is an extension of the above uses of the LFSR into a general test block which can be reconfigured into a number of different operations. The register consists of a series of identical cells which comprise a D-type flip–flop and associated logic. The basic cell is illustrated in Figure 6.14. As

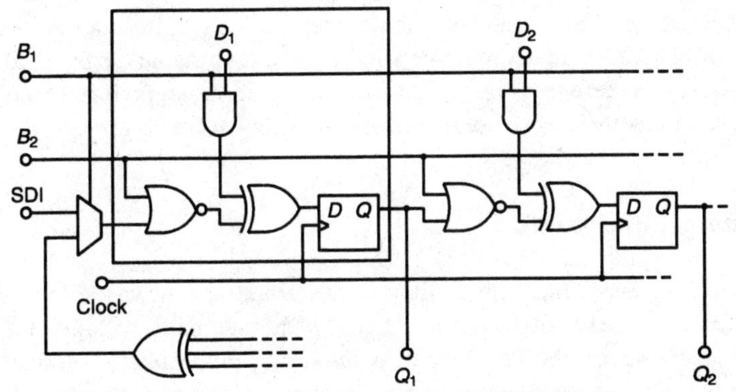

Figure 6.14 Configuration of BILBO cell

well as the basic input and output to the flip–flop, the cell also has a scan data in (SDI) input and two controlling lines, B_1 and B_2. The combination of these control lines means the register can be configured into four modes of operation:

$B_1 = 0 \qquad B_2 = 1$

The exclusive-OR gate on the flip–flop input has both inputs set to 0, so the input to each D-type is 0. After the next clock pulse, all stages of the register will be 0, so this mode represents a synchronous reset.

$B_1 = 1 \qquad B_2 = 1$

The NOR gate is disabled and the AND gate is enabled (output follows D input). The links between the register stages are broken and the cells operate as individual flip–flops. The register is configured as a parallel-in, parallel-out system. This is the normal system operating mode, so no test functions are active.

$B_1 = 0 \qquad B_2 = 0$

The NOR gate is enabled and the AND gate is disabled. Each stage is connected to the next through an inverter. The data inputs are blocked, but the SDI is enabled through the multiplexer. The whole device operates as a shift register, so could be used for scan-path operations, SDO being available from the Q_n output.

$B_1 = 1 \qquad B_2 = 0$

Both the NOR and the AND gates are enabled. The fed-back signal is passed through the multiplexer to the input of the system and the whole circuit is configured as an LFSR. There are now two possible operations: with the data inputs fed from observed nodes, the circuit will operate as a signature analyser; with the data inputs held at fixed values, the circuit will operate as a PRBS generator. The taps are chosen to operate as a maximal-length sequence generator. The data inputs must be such as to avoid the all-zeros condition in the register. As there is an inversion stage in the input logic, it is acceptable to set the D inputs to 0, as this will pre-load the register with all 1s.

6.4 Supply current testing

In section 6.3 the 'stuck-on' fault was described and it was shown that the fault was extremely difficult to test using the standard voltage test vector approach. However, the presence of a stuck-on fault is likely to mean that at certain operating states there will be a low resistance connected between the power supply and the ground. Therefore a relatively large supply current will be drawn. This is particularly easy to spot in the case of CMOS, where the normal quiescent current drawn is virtually zero. Simply by monitoring the amount of quiescent supply current drawn by the IC (IDDq) we should be able to detect the stuck-on faults. In addition, IDDq testing can also be used to detect such faults as bridging and gate oxide short faults, which are also difficult to test by the normal voltage test vector approach. In general, IDDq will help in detecting faults based on short-circuit effects where higher than usual currents may flow. IDDq has difficulty in detecting open-circuit-based faults, unless those faults propagate in such a way as to cause high-current faults further down the logic chain.

So, in general, IDDq testing must be done as a supplement to the normal test vector measurement. It can of course be done at the same time. The process is to have a current monitoring circuit placed in series with the IC power or ground supply line. Although the latter should strictly be referred to as ISSq, the term IDDq has been adopted for all forms of supply current test. Under the control of automatic test equipment (ATE), the normal set of voltage test vectors is applied and the outputs monitored. For any or all of these test vector states, the supply current can be monitored and recorded. There is a problem, however, in that the current monitoring operation is in general a much slower one than the voltage measurement and to IDDq test at every vector condition would cause the overall test to become excessively long. Hence, usually only a subset of test vectors has an associated IDDq measurement, selected by the engineer or the ATPG software, such that it is likely to test for the other fault model types. Additionally, a set of 'supplementary' test vector conditions may be used which are specifically for setting up IDDq measurement conditions and would not normally be part of the voltage test vector set.

6.4.1 Practical implementation of IDDq

At first sight it is a very simple operation to implement a practical IDDq measurement system, but as with most things in engineering there are practical difficulties and trade-offs between approaches. Some commercial ATE hardware can directly monitor the current flows in its external leads. However, the line being monitored must be protected from voltage spikes from the ATE meter when it changes ranges. So a capacitor must be included to suppress these spikes, which means that there is a settling time associated with the measurement, typically up to 40 ms, slowing the rate of measurements to 25 per second at best. Therefore only a very small subset of test vectors can be associated with the IDDq test, or alternatively a very long test time must be accepted. In addition, the resolution of such a current measurement is not very satisfactory, hence small changes and small values of current cannot be reliably detected.

The alternative is to have a current monitoring circuit that converts the current flow into a voltage which may be more accurately monitored. The simplest realization would be a resistor in series with the supply line across which a voltage would be developed that is proportional to the current flow. However, as the current varies, the supply voltage to the IC also varies, and this can cause operating difficulties. Another approach is to use a form of current mirror that provides a more buffered loading on the supply line. Such a circuit may be formed on the IC itself, so there are two approaches: on-chip and off-chip IDDq testing. Each has its own advantages and disadvantages. On-chip circuits tend to operate faster but still fit in with standard ATE. They do have a cost in terms of chip overhead, and care must be taken in circuit layout. Off-chip circuits tend to be slower but have no circuit overhead, and this is the likely approach to be recommended as measurement standard by a recently formed Quality Test Action Group (QTAG).

6.5 Analogue test

The preceding discussion in this chapter has been solely about digital testing. Analogue testing developed, largely, independently and has a very much different basis. There are many reasons for this. The nature of the two types of circuit is fundamentally different, the types of signals being dealt with have different properties and the faults being tested for are different. The usual nature of digital testing is to apply a series of input signals and to monitor a series of 1s and 0s from one or more outputs. These are checked against an expected 'correct' set of values; if any of the bits differ then the device under test is deemed to have failed and the test *need* not continue. Such testing is termed *deterministic* as the expected output has been completely determined before the test starts. With analogue circuits, by their very nature, there is an allowable variation in the signal parameters, owing to process variations, within which the device can still be deemed to be functional. The output may not agree precisely with the expected result, and the whole test must be completed and some assessment made of the complete set of signals before a judgement of pass or fail can be made. Such testing is called *non-deterministic*.

Another major difference between the two types of testing is that in digital testing we are concerned mainly with the measurement of a voltage signal as a function of time. As has been seen, current testing is also sometimes employed. In analogue testing, a number of different types of signals may be measured as a function of different parameters. Thus voltage, current, gains, impedance, noise, harmonic content and other non-linear effects may all be considered, and time, frequency, temperature and other parameters may be used as functional variables. So the range of possible tests is enormous, and it depends on the particular analogue circuit function which of the tests are to be adopted.

6.5.1 Analogue fault modelling

There are similarities and differences between digital and analogue fault modelling approaches. Both have their origins in process defects at the fabrication stage and some models cover both types of circuits; they are used in different ways and in analogue there is a further class of fault model.

The application of fault models in digital circuits is performed at the gate level where the building blocks implement Boolean functions and the faults can be modelled in terms of these functions, so generating such models as the stuck-at fault. Fault models in analogue are usually applied at the transistor level, for implementation in simulators such as SPICE, which is at a lower level of abstraction than in digital. There is no equivalent of the stuck-at fault in analogue fault modelling, the open and short circuits that give rise to such faults must be modelled as such at the transistor level. Such faults are termed *catastrophic* faults, as they usually lead to the non-functioning of the transistor that they are associated with. They are also termed *hard* faults in order to discriminate them from the other type of analogue faults which are termed *soft* faults. These occur as a result of parameter variations, such as the values of resistance, capacitance or transistor gain, which cause the overall circuit to operate outside its acceptable limits. In other words, the fault does not cause a complete non-functioning of the circuit, but one or more of the functional specifications is beyond a pre-defined range.

The open- and short-circuit hard faults associated with a transistor are easily incorporated into a fault model for simulation purposes. However, problems may arise in some simulators if the nodes are directly open or short circuited. Instead, high- or low-value resistors are substituted to model the open and short circuits respectively. In addition, capacitors are often included to model the capacitive effects of open circuits. A generalized hard fault model for a MOS transistor which is in common use is illustrated in Figure 6.15. Bridging faults can be included in a similar way to those in digital cir-

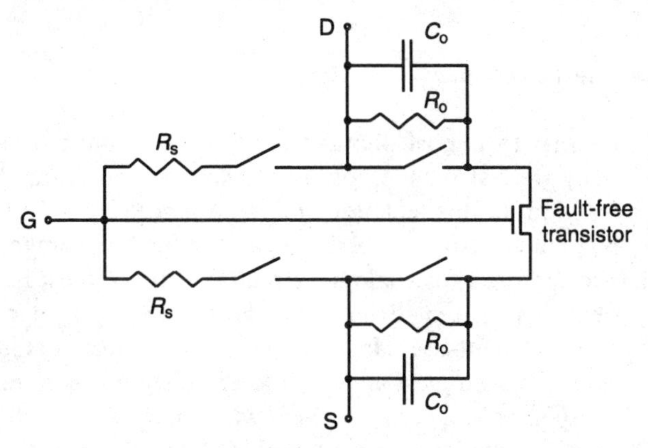

Figure 6.15 Generalized hard fault model for FET

cuits, with the same problems of requiring a knowledge of the IC layout before the likely bridging faults can be estimated. The GOS results in a very complex fault model, as illustrated in Figure 6.16. Here the short circuit divides the transistor into two separate devices with lengths dependent on the actual position of the short. In addition, the connection from the polysilicon gate to the channel forms a rectifying contact modelled by the diode and resistor.

Soft faults are particularly difficult to estimate, as any particular passive device can take on an infinite number of values within a particular process, given the associated variations. Some research work has been done to study statistically the process variations to see the effect this has on the various component tolerance values, but this information is used more often in the design process than as a testing tool.

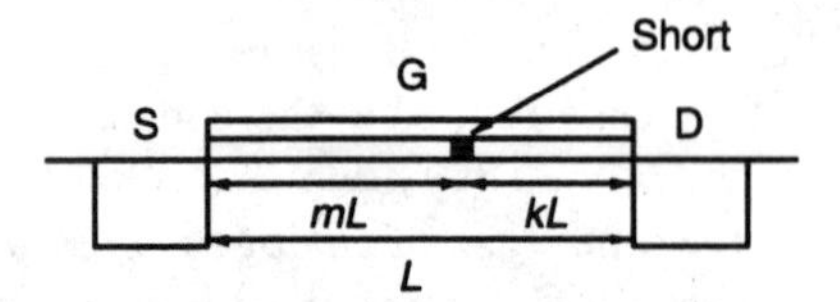

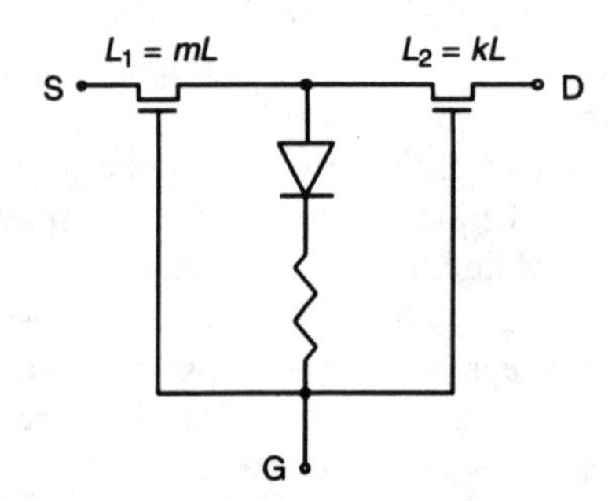

Figure 6.16 Gate oxide short and its transistor-level fault model

6.5.2 Analogue test strategies

We have seen earlier that most digital testing is based on a structured test approach, whereby we test for the possible faults that can occur and not the actual circuit function. In this technique the fault model is used to determine the necessary input test vectors to obtain the highest fault coverage. Using the non-deterministic approach for analogue circuits, structured test is not practical as the concept of path sensitization to transfer the result of a test to an observable node is not realizable. Instead, the basis of most analogue testing is a functional one. This simplest approach to analogue test is to measure the primary output responses and compare them with the expected output derived from simulation or a known good circuit. Although this is relatively quick and easy, there are a number of drawbacks. First there will be a tolerance band around the expected signal values in which a circuit may still be deemed to be operating correctly. The size of this tolerance band will depend on a number of factors including the fabrication process used, the circuit function and its use in a larger circuit or system. In addition, circuits for military or safety-critical use may have tighter specifications than those for commercial circuits.

More complex testing strategies have been devised, making use of the fault models to provide a degree of diagnosis or DFT. One approach is based on a fault dictionary whereby a series of simulations of the circuit is performed, introducing each expected fault in turn, and recording the result. This builds up a dictionary of faulty responses, and from this a measure of fault coverage can immediately be derived, as not all faults may cause the circuit to operate outside its acceptable limits. The circuit may then be tested and if the output is outside the acceptable range, some form of fault diagnosis can be performed by comparing the output signal with the entries of the fault dictionary. It may of course be that some faults give a similar (erroneous) response, so it may not be possible to diagnose the position of the fault precisely. There are other approaches whereby the circuit is examined in detail to estimate the number of faults that can be detected given particular input tests and test points. This can provide information as to the nodes that should be made directly observable in order to maximize the testability of the circuit. These techniques have largely been developed using discrete circuits where access to circuit nodes is relatively easy. There are problems in accessing nodes on ICs and these techniques have not yet been widely adopted for IC test.

6.6 Mixed-signal test

As has been noted in previous chapters, the availability of reliable CMOS and BiCMOS processes has enabled ICs to be realized with both analogue and digital blocks of circuitry on the same chip. There are difficult problems involved in testing such mixed-signal ICs, owing to the different development of digital and analogue tests. A generalized mixed-signal IC is illustrated in Figure 6.17, indicating that the IC may have both analogue and digital primary inputs and outputs, while there are a number of possible interfaces internally between the blocks.

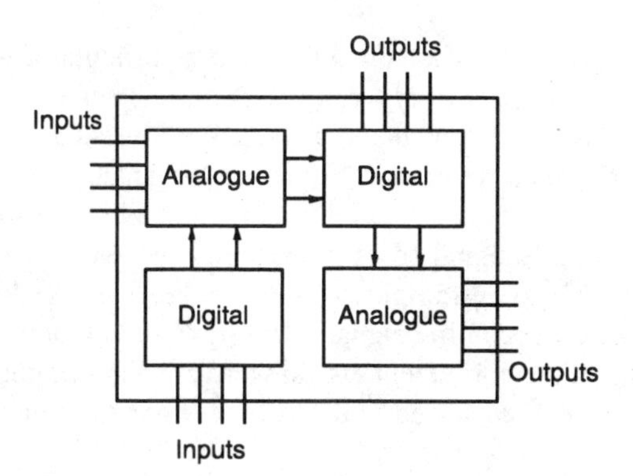

Figure 6.17 Generalized mixed-signal IC

The only current commercially viable approach to mixed-signal testing is physical partitioning of the circuit into the separate analogue and digital blocks at the interfaces, with mode-specific testing then being applied to each block. There are obviously many drawbacks with this approach. Primarily there is an overhead cost in providing access to the interface nodes to make them observable. Secondly, two sets of test equipment, test designs and operations must be made. There are also problems in separate testing of the blocks, as timing relationships of signals in different blocks may have an effect on the overall functionality of the IC, and these relationships cannot be checked if the blocks are testing in isolation.

6.6.1 DSP emulation of analogue test equipment

Techniques have been developed, however, that largely eliminate the latter of these problems. These are based on the use of digital signal processing (DSP) to emulate the analogue test hardware, such as signal sources and voltmeters, in software. This is more than just controlling the test equipment by a computer; the test equipment does not physically exist – interfaces are established using DACs and ADCs and all of the equipment functions are generated in software. For example, an analogue waveform as a function of time can be generated as illustrated in Figure 6.18. The digital data sequence, which con-

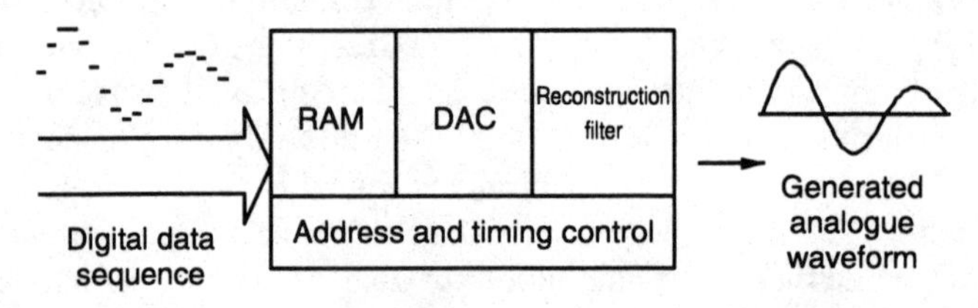

Figure 6.18 DSP emulation of waveform generator

sists of a word representing the voltage level at a particular time point, is read into the RAM. Under the correct timing control, this sequence is then passed into the DAC and the resulting analogue signal is filtered to obtain the correct function which can then be used to stimulate the analogue block under test. Likewise the reverse process of monitoring a signal is illustrated in Figure 6.19. Here the analogue signal being monitored is passed through an anti-alias filter (as the signal is to be sampled at a particular rate, this limits the allowable bandwidth of the incoming signal). The signal is sampled and passed into an ADC, generating the equivalent digital word at each sampled point in time. This data sequence is then stored in RAM for further examination or processing.

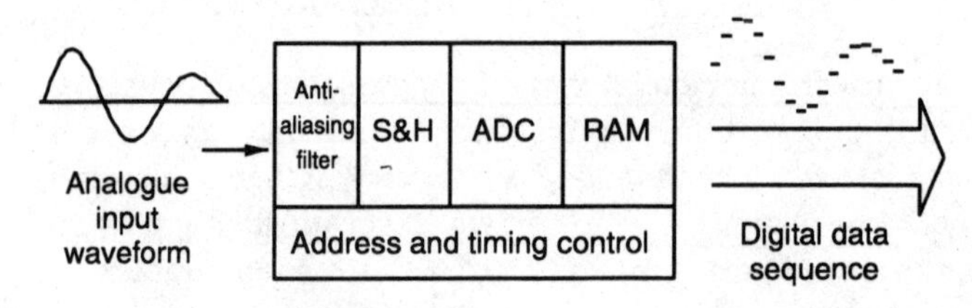

Figure 6.19 DSP emulation of waveform digitizer

Using this approach, the testing of mixed-signal circuits can be performed under the central control of one computer with a synchronized timing system, so enabling correct time relationships of signals in different blocks to be maintained. It still does not get around the problems of physical partitioning and the cost of test equipment (often of the order of $1 million) makes this solution unavailable to small commercial enterprises or academic laboratories.

6.6.2 Other approaches to mixed-signal test

While it has been noted that physical partitioning and mode-specific testing is the only practical approach to mixed-signal testing at present, a number of other techniques have been proposed in an attempt to obtain a truly 'unified' approach and eliminate, or at least reduce, the need for partitioning. Although many of these are still at the research level, some may be adopted in the future.

One approach that has already shown great potential is the IDDq supply current approach. This is further helped by the fact that most mixed-signal ICs have separate power supply lines for the digital and analogue blocks. This is because the sharp signal transitions in the digital circuits can induce spikes in the power supply which would severely affect the operation of the more noise-sensitive analogue circuits if they were supplied by the same rail. So there is a built-in partitioning system with no extra overhead incurred. It has already been shown that IDDq testing can identify faults not picked up by other voltage-based approaches. At this stage, current testing is only used as a supplement to voltage testing, although this may change in the future.

As the digital techniques and testing approaches are more formalized and in general more developed than analogue testing, an obvious approach to mixed-signal test would be to model the analogue elements in terms of digital equivalents, to which standard digital test approaches can then be applied. Clearly any digital model of an analogue circuit will be a fairly crude approximation, owing to the different nature of the signals involved, but this technique may be of some use in testing hard faults occurring in the circuit. As an example of this approach, a digital equivalent of an analogue comparator circuit is illustrated in Figure 6.20.

A more formal approach to this technique is termed the functional K-map technique. This algorithm consists of a number of stages:

(1) The analogue circuit is simulated under fault-free conditions and the response to certain input signals is calculated.
(2) This operation is translated into a Karnaugh (K)-map. In order to do this, a threshold voltage must be defined to divide the analogue voltages into 1s or 0s.
(3) The digital function(s) defined in the K-map and the equivalent digital circuits to realize these functions are derived.

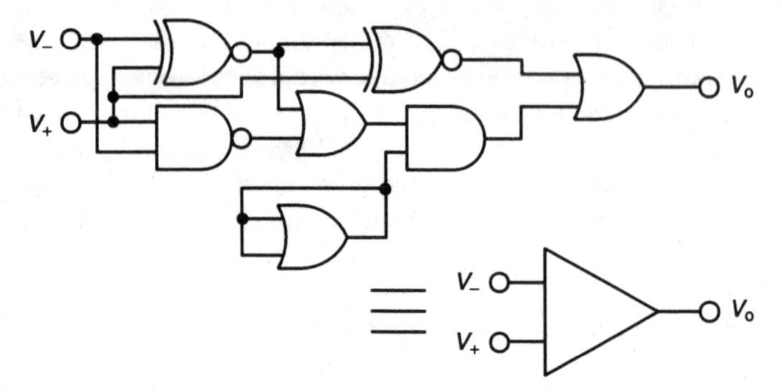

Figure 6.20 Digital equivalent circuit of analogue comparator (Brown and Damianos, 1983)

(4) These circuits are incorporated into the digital circuits and an overall testing strategy devised.

While these are truly unifying approaches to mixed-signal test, the equivalent circuits are very poor models and the fault coverage is likely to be low, so it is unlikely that these approaches will ever be developed into a practical technique.

If the approach of unifying the circuit models is perhaps impractical, the next technique considers the unifying of the testing signals. If, for example, an analogue circuit is stimulated by a pulse, or series of pulses, which is digital-like in nature, the response of the circuit will consist of a transient signal. This signal should be characteristic of the circuit under test, that is any faults within the circuit should change the transient response. Additionally, the output signal should be (largely) digital-like in nature, and so both the stimulus and resulting signals should be capable of being propagated through intervening digital blocks of circuitry. Such a technique of Transient Response Testing (Evans *et al.*, 1990) has been proposed as a unifying approach to mixed-signal testing. Some promising results have been found for small-scale circuits and good fault coverage obtained without the need for partitioning. However, its effectiveness in larger scale circuits, where the test signals may have to be propagated through several stages with resulting loss of test information, has yet to be proved and the technique has not yet been widely adopted.

References

Brown, D. and Damianos, J. (1983) Method for simulation and testing of analogue/digital circuits, *IBM Tech. Disclosure Bulletin*, Vol. 125, pp. 6367–6368.

Evans, P.S.A., Al-Qutayri, M.A. and Shepherd, P.R. (1990) On the development of transient response testing for mixed-mode ICs, *Journal of Semicustom ICs*, Vol. 8, No. 2, December, pp. 34–39.

Roth, J.P., Bouricius, W.G. and Schneider, P.R. (1967) Programmed algorithms to compute tests to detect and distinguish between failures in logic circuits, *IEEE Transactions on Electronic Computers*, Vol. EC-16, No. 5, October, pp. 567–580.

Bibliography

P.H. Bardell, W.H. McAnney and J. Savir, *Built-In Test for VLSI*, Wiley, New York, 1987.
T.E. Dillinger, *VLSI Engineering*, Prentice-Hall, Englewood Cliffs, New Jersey, 1988.
P.J. Hicks, *Semi-Custom IC Design and VLSI*, Peter Peregrinus, London, 1983.
F.J. Hill and G.R. Peterson, *Computer Aided Logical Design with Emphasis on VLSI*, 4th edn, Wiley, New York, 1993.
R.E. Massara, *Design and Test Techniques for VLSI and WSI Circuits*, Peter Peregrinus, London, 1989.
K.P. Parker, *The Boundary Scan Handbook*, Kluwer Academic Publishers, Norwood, Connecticut, 1992.
D.A. Pucknell and K. Eshragian, *Basic VLSI Design*, 3rd edn, Prentice-Hall, Sydney, Australia, 1994.
R. Rajsuman, *Digital Hardware Testing*, Artech House, Boston, Massachusetts, 1992.
G. Russell and I.L. Sayers, *Advanced Simulation and Test Methodologies for VLSI Design*, Chapman & Hall, London, 1989.
R.S. Soin, F. Maloberti and J. Franca, *Analogue Digital ASICs*, Peter Peregrinus, London, 1991.
B.R. Wilkins, *Testing Digital Circuits – An Introduction*, Chapman & Hall, London, 1986.

Problems

6.1. For the CMOS inverter circuit, what is the appropriate stuck-at fault model to apply if
(a) the source and gate of the p-channel FET are shorted together?
(b) the drain of the p-channel FET is open circuit?

6.2. How many tests are needed to detect all the possible single stuck-at faults in a 3-input NAND gate? List the test patterns and the faults that they cover. [4]

6.3. For the circuit of Figure 6.2, apply the D-algorithm to derive a test pattern to detect a stuck-at-1 fault at point *G*. [1101]

6.4. If in Figure 6.2, the lower input to the final AND gate was connected to point *G* instead of input *D*, is the fault *E* s-a-0 still testable? If so, what is the test input? If not, why not?

6.5. For the circuit of Figure 6.3, apply the Boolean difference algorithm to determine the complete set of tests for stuck-at faults on input line *C*. Which tests detect which fault? [$\overline{X}_1\overline{X}_2X_3$, $X_1\overline{X}_2X_3$, $X_1\overline{X}_2\overline{X}_3$, $\overline{X}_1\overline{X}_2\overline{X}_3$]

7 Afterword
The future

7.1 Introduction

The subject matter of this book is one of the fastest moving technologies in the field of Electronic and Electrical Engineering. While this has been written as an introduction, and much of the basic material remains relevant, certain aspects can become out-of-date fairly quickly. Until later, updated, editions of the book appear, the author will try to alleviate this problem by attempting a very brave task – predicting the future.

In reality the task is slightly easier than may be first thought: many trends are identifiable in terms of speed, size of circuit, etc. In addition there are new technologies in the research stage that could well appear as production tools in the near future. This chapter will contain a mix of these, in order to make an attempt to paint a picture of IC design and manufacture in 5 to 10 years' time.

7.2 Fabrication technology

7.2.1 Dimension reductions

Much of the improvements in terms of circuit area (and hence packing density), wafer size and circuit speed have been directly driven by improvements in semiconductor processing and IC fabrication technology. The advantages of smaller ICs (or more circuitry in a given area) and faster speed are obvious. The main parameter that controls such changes is the minimum dimension (λ) that can be defined by the photolithographic process. This in turn is limited by the chemical resist that is used to protect defined areas from the etching (or lift-off) process. There is a natural limit in terms of the wavelength of the activating beam. For ultra-violet (u-v) light this is a few tenths of a micron, which is the current dimension limit for production processes: 0.5–0.6 micron processes are readily available from a number of foundries, 0.35 micron processes are to be introduced very shortly. Plans are being made for a 0.25 micron process within the next three years, and a 0.1 micron process around the turn of the century.

These improvements can only be made by implementing deep u-v and X-ray lithographies, the wavelengths of which are shorter and hence have smaller dimensional resolution. Although the technologies are, or soon will be, available for use, this does not automatically mean that they will be used in a production environment, for this will still depend on the commerciality of making such a move to smaller dimension processes. The cost in new equipment, tooling, materials and training must be offset by improved profitability of the product(s). The decision as to exactly when any particular foundry makes a change to a new process can be a complex and difficult one. The certainty is that these size reductions will take place, the only unpredictable factor is exactly when.

7.2.2 Device speeds

With reduced dimensions comes higher speed of circuit operation. This is determined primarily by how fast current carriers (electrons and holes) can cross a device (transistor). The smaller the device can be made, the shorter the time, but this is not the only way in which circuit speed can be increased. Carrier velocity depends on the electric field strength and the carrier mobility; increasing either of these should lead to faster circuit operation. Increasing field can lead to problems of supply voltage and in any event the velocity saturates at high fields. The mobility is a natural parameter of a particular semiconductor and is hard to alter. Electrons have a higher mobility than holes in a particular semiconductor and are used wherever possible as the main current carrier (hence nMOS is a common technology, but pMOS only occurs in CMOS). So, to use a higher mobility, a different semiconductor must be used. This sparked an interest in GaAs as the material of the future, and great predictions were made in the 1980s of its increasing share of the IC market. Apart from its niche at microwave frequencies these predictions, again because of the high processing costs, have not been realized. Indeed, with the recent developments in personal communications at frequencies around 1 GHz, silicon is challenging GaAs for these high-frequency applications.

Germanium has higher mobilities than silicon, but presents problems in processing and circuit parameters and has effectively never been used in the IC market. Research is continuing into a mixed Ge–Si technology, making use of the advantages of both materials. Carbon has also been cited as a potential new IC material having a higher mobility. The technology problems involved in forming diamond substrates and their subsequent processing are enormous and it will be many years before this is available as a production technology.

7.2.3 Supply voltages

The influences of dimension scaling are many and varied. One in particular is that as the width of a conductor is reduced, the current density within the line increases, all other things being equal. High current densities can cause problems in terms of burn-out and electromigration in metal tracks. In addition, the resistance of lines is increased which can lead to greater power losses. Power consumption is of vital importance, particularly with the increase in portable equipment such as computers and telephones which are based on battery supplies. Many of these problems can be alleviated by reducing the supply voltage(s) to the ICs. The standard digital supply of 5 V is increasingly being supplanted by a 3.3 V supply. This is the preferred supply voltage for the submicron circuits and is likely to dominate within 5 to 10 years. As λ values shrink to 0.1 micron and below, further reductions in supply voltage are to be expected, already research work is being undertaken with 2.5 V supplies.

These reductions of supply voltage bring their own problems of course. The absolute noise margin values are correspondingly reduced, and circuits can be more prone to interference. The threshold voltage for MOS devices tends to scale with supply voltage. The V_T for an nMOS enhancement device is typical $0.2V_{DD}$. In bipolar circuits, however, the active V_{BE} is material dependent and is typical 0.7 V for silicon devices, independently of supply voltage. As supply voltages are reduced, the design constraints for bipolar circuits can become almost impossible to overcome. It is therefore unlikely that supply voltages for bipolar and BiCMOS circuits will be reduced much below 3.3 V in the future.

7.3 Technologies

As has been indicated elsewhere in this book (and is indeed the book's *raison d'être*), the growth in in-house designed, small-production-size ASICs has increased out of all proportion in recent years. It is expected that this trend will continue, albeit not at the same explosive rate. Much of this IC production is already in the form of gate arrays and it is expected that this form of architecture will dominate over the next few years. The reason for this is the very short design time associated with gate array architectures. There will be increasing moves away from mask programmable to field programmable devices. The difficulty of not being able to have field programmable sea-of-gate structures can be overcome by adding a further layer of metallization so that the microfuses can sit on an insulating layer above the actual gate array. A device based on this architecture has recently been introduced to the commercial market.

Because of the increasing portable applications for ICs, with their low power consumptions, it is likely that CMOS will become the dominant technology, although there will always be requirements for the bipolar and

BiCMOS technologies, particularly for the high-speed/high-power applications. Mixed-signal circuits will continue to grow, as again these are likely to be largely CMOS based, and will find their niche in the rapidly growing telecommunications markets.

GaAs technology has failed to reach the level predicted a few years ago. It will still find use for the very high frequency (> 2 GHz) applications, although silicon-based circuits are already encroaching into the GaAs domain. This trend is likely to continue unless the economic viability of GaAs is greatly improved.

7.4 CAD

Computers, manufactured as they are from ICs, have seen an equal, if not greater, expansion in their power and complexity. Computing tasks that 15 to 20 years ago could only be performed on large expensive mainframe machines and that then progressed 5 to 10 years ago to workstations, can now be handled by personal computers. As many of the CAD tools associated with IC design and manufacture, such as simulation, layout and error rule checking, are very CPU and memory intensive, these were, until recently, beyond the scope of most computers. That situation has now changed and very sophisticated IC designs, including all the computer-aided stages, can now be performed on relatively modest PCs.

The growth in power and capability of computers shows no sign of diminishing, at least over the next 5 to 10 years, and the performance may have increased by orders of magnitude within this time span. With this increase in power, there will no doubt be an increase in the capabilities of the CAD software packages for IC design. This will not only be in the realms of increased processing speed and memory capabilities, but also in the sophistication of the tools. There are a number of areas in IC design that will particularly be able to benefit from these increases in power.

Firstly, we shall consider simulation, which can be one of the most computer-intensive tasks. As was mentioned in the previous chapters, the accuracy of simulation depends on two factors – the accuracy of the mathematical routines applied by the simulator, and the fidelity of the device models. Both of these can be improved with greater power – more sophisticated simulation routines can be devised, which take more processing, and more complex circuit and device models used, which better reflect the real device behaviour. The areas of analogue and mixed-signal simulation have more scope to benefit here. Analogue simulation tends to be performed at a lower level of abstraction than digital, and so the same size of circuit (in terms of transistor numbers) will take longer to simulate if the circuit is analogue rather than digital. With mixed-signal circuits there is an added complication in the different signal descriptions of the two modes, and the way the simulator handles these at the interfaces. Mixed-signal circuits can be, and have been, simulated totally at the transistor level, effectively treating the digital circuits in an analogue manner with continuously variable signal levels. Simulation at this level, even today, is very time consuming, and is only practical for circuits up to MSI. In future, large-scale circuits will be possible, and true mixed-signal simulators, which are currently rare, will be much more common.

Design for testability is another area that will particularly benefit from improved computing power. Concurrent circuit diagnosis and feedback on testability to the designer at an early stage will become more important, particularly in the full-custom approaches. The algorithms for such tools are very sophisticated, and largely only at the research stage at the moment. Increased computing power will make the tools commercially useful in the next 5 to 10 years.

7.5 Testing

The reductions in feature size, leading to increased density of components on an IC, have a fundamental effect on the testing process. Given a linear reduction in the feature size, the complexity (number of components on chip) increases as a square function, while the number of bond pads around the edge can only increase linearly. So there are fewer nodes per component available, and controllability and observability, the two important parameters of testability, are decreased. Add to this problem the increase in mixed-signal circuits, and the overall testing problem can only increase in the future.

Faced with these problems, new techniques must be introduced to solve them. Advances have already been made in design for testability, the inclusion of built-in self-test and scan-path approaches. All of these techniques help the problem of naturally decreasing testability with decreasing feature size. The 'cost' of overhead must be viewed more positively as the circuits will have to include dedicated test circuitry in order to maintain product quality. Scan-path techniques will become increasingly important as the IEEE 1149.X standards are introduced over the next few years. Analogue scan paths will also have to be developed for use in the mixed-signal circuits. The general mixed-signal test problem, in terms of physical partitioning of the circuit and mode-specific testing, is still to be solved. Some 'unifying' techniques are being researched, but none is yet commercially viable, although this may well alter in the next few years.

One technique that is imminent is the concept of the 'virtual test bench'. Virtual design techniques are based on the computer simulation of complex systems, making use of models of various electronic instrumentation connected together. The performance of such systems can be assessed before the system is finally constructed. These simulations can also interface with controllable instrumentation as required, providing a mixed system of hardware and software. Such virtual design systems are already commercially available. The advance to the virtual test bench is only a small one. Here the emphasis will be on an assessment of test techniques which will be applied to an IC design before the chip is actually manufactured. Problems that would not normally manifest themselves until this stage can be identified, saving a design iteration.

7.6 Conclusion

The afterword supplied by this chapter has provided a brief look at some of the likely developments in IC realization over the next few years. One certainty is that the range, complexity and power of ICs will continue to increase at the rapid pace that they have over the last 30 years. A few of the new techniques and approaches that will fuel these increases have been described here, based on current trends and research work in the area. It is possible that completely new techniques, materials and ideas will emerge in the future which could completely transform the state of the art. To be able to predict such revolutions is beyond the power of the author however!

Appendix 1: The Fabrication Process

It is not essential that an engineer has an intimate knowledge of the IC fabrication processes in order to design a circuit successfully. Much of the layout process that generates the data the fabrication house uses is automated. However, it is useful to have a basic knowledge of the fabrication process in order to appreciate some of the design techniques described in this book. In addition, the faults being tested for have their origins in the fabrication of the chip. Therefore the information on the fabrication process has been included in this appendix, with the strong recommendation that engineers with little or no knowledge of IC fabrication processes read this before the other chapters of the book in order to get a full appreciation of the design and test techniques described.

A1.1 Semiconductors

As their name suggests, semiconductors have electrical properties between that of insulators (such as ceramics and plastics) and conductors (such as metals). Indeed, from a material point of view they are closer to insulators. For a material to be able to conduct an electric current it must have mobile charge carriers (electrons) available. The electrons orbit the atomic nuclei in a crystal structure in a series of energy bands. Each band can only contain a fixed number of electrons and the atoms bond together in such a way that normally all the energy bands are full. The highest filled band, in energy terms, is called the valence band. The next available band, at a higher energy level, is called the conduction band. The electrons in a full band are not capable of moving through the band (there are no 'spaces' to move into), so for a material to conduct an electric current there must somehow be some electrons in a partially filled band, normally the conduction band.

There are two ways that this can come about. In normally insulating materials there is a relatively large energy difference between the valence and conduction bands. To get an electron to transfer from the normally full valence band to the normally empty conduction band it must be given sufficient energy. This usually comes from thermal energy (that is, by raising the temperature), but may also be derived from light or other radiation energy (the basis of opto-electronics). In natural conductors, the band structure of the material is such that the two bands naturally overlap, so even when the valence band is full, the conduction band appears partially filled and conduction can take place. Semiconductors have a band structure like insulators, with an energy gap between the two bands. However the bands are relatively close in energy terms and at room temperature sufficient electrons have received energy and transferred into the conduction band for the material to have a significant conductivity.

There are two disadvantages to the conductivity associated with these pure, intrinsic semiconductors. Firstly, although significant, the conductivity is still fairly low, typically 8 to 10 orders of magnitude less than a metal. Secondly, as the conducting electrons are generated by a thermal process, the conductivity is highly dependent on temperature. This would mean that any circuits based on these materials would have characteristics that varied greatly with temperature. It is much more preferable to have circuits whose performance varies very little with temperature.

There is a solution to both of these problems – the introduction of a small amount of 'impurity' material, that is a different material from the basic semiconductor. These come in two sorts. The first is those materials that have more electrons in the outer bands than the semiconductor and so readily give up electrons for conduction. Since they donate electrons to the system, they are termed donor impurities and the resulting extrinsic material is termed n-type. Alternatively, material may be introduced that lacks electrons in the outer band and accepts electrons from the semiconductor, leaving a shortage in the valence band. Hence the partially filled band is now the valence band. As the conduction process is slightly different in this band, it is convenient to consider that the current is conducted by the vacancies or 'holes' which are considered to be particles in their own right with a positive charge, equal but opposite to that of an electron. As these impurities receive electrons from the system, they are termed acceptors and this leads to p-type semiconductors.

Whichever form of impurity, 'dopant' material, is used, the conductivity of the natural semiconductor can be increased by several orders of magnitude with only very small amounts of dopant, typically one part per million. Also, as the process of donating or accepting electrons requires only a very small ionization energy, at room temperature all the electron exchanges have taken place and the conductivity of the material hardly varies with temperature.

A few elements are natural semiconductors, the primary ones being silicon (Si) and germanium (Ge). Other semiconducting materials may be formed from a mix of materials. Usually this involves an element from group III of the Periodic Table mixed with one from group V, or sometimes a mix of group II and group VI elements, leading to III–V compound semiconductors (such as gallium arsenide (GaAs), or indium phosphide (InP)) and II–VI compound semiconductors (such as cadmium telluride (CdTe)).

A.1.2 Material preparation

As the level of dopants in a semiconductor is extremely low, all the operations when processing materials and devices have to be done in extremely clean conditions to eliminate as far as possible contamination from dust and other particles. Integrated circuits are processed in the form of thin circular wafers in which the processing operations are largely performed on the top surface of this wafer, termed a 'planar' process. So the first stage of processing is to transform the basic semiconductor material into the usable wafers.

The two restrictions on the processed material are that it must be free of unwanted contaminants, only the semiconductor and any required dopant material being present; and also the material must be in a very regular crystal structure. Thus the initial preparation has to be done in two stages – purification and crystal growth.

If a semiconductor is molten and then allowed to solidify, it will normally do so as a polycrystalline structure, that is there is no regular crystal structure throughout the block of material, there are simply small areas of crystals, at different orientations to neighbouring crystal areas. It is while the material is in this state that the purification process is performed. The basic approach is illustrated in Figure A1.1. The bar of material is selectively made molten by a radio-frequency induction heater which is slowly moved up and down its

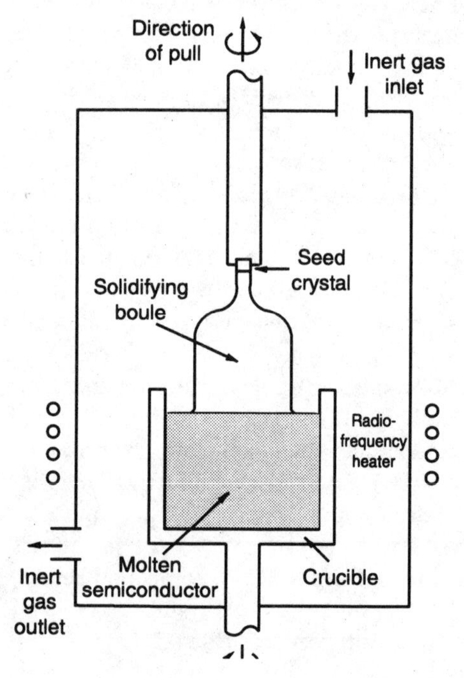

Figure A1.1 Zone refining process

length. Any impurities within the material tend to remain within the moving molten zone. After a series of passes, virtually all the unwanted material is confined within the molten zone which is then guided to one end, removed from the main bar and discarded.

Once the material has been purified, there are two main techniques for producing the single crystal bar; both techniques require a seed crystal of the correct orientation. These two techniques are the Czochralski method and the float zone method.

The Czochralski method is illustrated in Figure A1.2. The purified material is completely molten, and the seed crystal is dipped into the surface of the melt and slowly withdrawn and rotated. The speed of pull and the rate of cooling will determine the diameter of the final rod or boule of material. Dopant material can be introduced into the melt in the required ratio.

The float zone method is very similar to the purification process illustrated in Figure A1.1. In this case the seed crystal is introduced at one end of the bar and the molten zone transferred very slowly from that end to the other. As the back end of the zone solidifies it will follow the crystal structure of the seed crystal, so the bar will eventually become a single crystal structure. Dopant material can be introduced in the form of a gas which is absorbed into the molten region of the crystal.

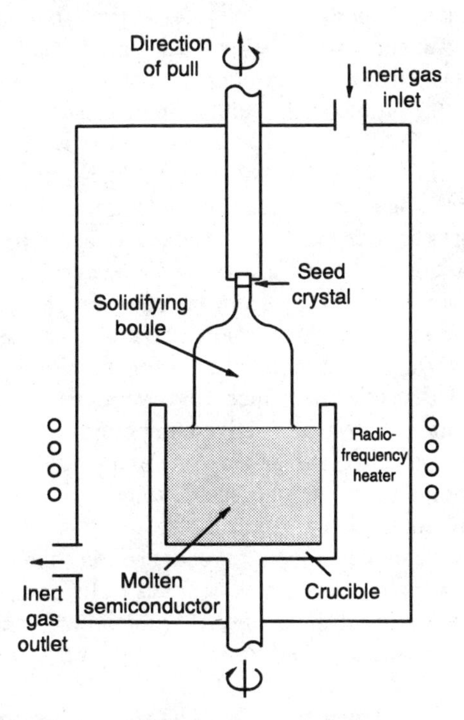

Figure A1.2 Czochralski crystallization process

In both cases the process takes a long time, as the cooling must be very gradual to achieve a consistent crystal structure. Typically it takes several hours to produce a boule of material.

Once the boule has cooled to room temperature, the wafers are sliced using a mechanical saw. This inflicts much mechanical damage on the surface of the wafer, so this must then be polished and finally etched to produce a mirror smooth finish that can be subsequently processed.

A1.2.1 Epitaxial growth

Although a certain amount of dopant material can be introduced into the material during the crystallization process, there is a limit on the level of dopant that can be introduced if a consistent dopant concentration is to be maintained throughout the material. In general, only a 'background' dopant level is introduced at this stage. To get the highly doped regions required by some of the active devices, a further crystal growth process is performed, providing a thin epitaxial region on top of the basic wafer. Again the consistent crystal structure has to be maintained so that the resulting wafer is still a single crystal throughout.

There are a number of approaches to epitaxial growth. Liquid phase epitaxy (LPE) involves the substrate wafer being dipped into a melt consisting of more semiconductor with a large amount of dopant material. Only thin epitaxial layers are required, so the problems of maintaining consistent levels of dopant are no longer relevant.

Vapour phase epitaxy (VPE) involves introducing two or more reactive gases to the substrate, one based on the semiconductor material and one based on the dopant. The two gases react in a heated chamber to form the solid semiconductor with the dopant, which grows on the heated substrate crystal. Any by-product gases are vented away from the reaction chamber. For example, silicon tetrachloride in the vapour form can be introduced with hydrogen and these will react to form solid silicon with hydrogen chloride gas as a by-product. Dopant materials can be introduced in a vapour form at the same time, either as an elemental gas or as a reactive compound.

Molecular beam epitaxy (MBE) takes place in a very high vacuum chamber and consists of firing a beam of constituent atoms at the substrate where they stick and form the continuing crystal structure. To introduce dopant, or to form compound semiconductors, several beam sources must be available. The process is very slow, as the vacuum system has to be pumped down between each operation, but control of the supply of epitaxial material is very good, so very precise doping profiles or very thin layers can be achieved by this technique.

Metal organic chemical vapour deposition (MOCVD) is similar to the VPE process in that the semiconductor materials and dopants are introduced as gases, in this case as organic-based compounds. The difference is that in the MOCVD process the reaction chamber is kept at room temperature, only the substrate is selectively heated to initiate crystal growth. As the reaction is relatively slow, like MBE this process is suitable for precise doping profiles and very thin layers.

A1.3 Photolithography

To construct devices and circuits on the surface of the semiconductor wafer in the planar process requires the transfer of the layout patterns on to the wafer. The patterns describe where areas of further doping, insulating layers, metal interconnections, etc. are located. To build up a functioning circuit may require the definition of many such layers on top of each other. The layer patterns are usually computer generated, based on the position and interconnection of the basic building blocks in the circuit, and consist (mostly) of polygonal shapes. The alignment of one layer on top of the other is of crucial importance to the success of the circuit fabrication. Knowing the tolerances associated with the fabrication processes, layout rules are available for any particular process, and these specify the closest allowable distances between the edges of different layers.

The technique of transferring the layer patterns on to the wafer is based on having layers of resistive material covering the surface, areas of which can be selectively removed to expose the circuit below to doping, etching or deposition of further layers. The process has many similarities to photographic techniques and is termed photolithography.

A1.3.1 Basic photolithographic process

In order to transfer the layer patterns on to the wafer, the first step is to transfer each pattern on to an actual-size plate that consists of areas that are transparent and opaque. These plates are called masks and usually consist of a glass plate with the pattern defined in a metal layer, often chrome. Where the metal is present the area is obviously opaque, and where there is no metal the mask is transparent.

The next stage is to cover the entire wafer with a chemical layer in which the pattern will be transferred from the mask. The chemical is sensitive to light and is called photoresist; it is an organic material with chains of molecules. When exposed to light the resist either polymerizes, making even longer chains of molecules and hardening, or the molecules break up and the material softens. These two types of resist are termed negative and positive respectively, and require inverse mask patterns to provide the same pattern of resistive layer. Whether positive or negative, the photoresist comes in a liquid form. The application of a thin film of resist is achieved by placing a small drop on top of the wafer and then spinning the wafer at high speed (typically 4000 to 6000 r.p.m.). This results in a very thin, even layer of liquid resist, the thickness of which is inversely proportional to the spin speed. The resist must be allowed to dry and solidify, and in the case of positive resist, oven baked to ensure that it is fully polymerized prior to exposure.

After the application of light through the mask, the soft areas of resist have to be washed away in a 'development' procedure very similar to normal photographic techniques. This just leaves the polymerized areas of resist protecting certain areas of the wafer, with the other areas of the wafer exposed through the developed areas. As the resist is sensitive to light, particularly frequencies in the blue end of the spectrum, these photolithographic processes have to be done in an area lit by 'safe' yellow light.

In order that the remaining resist can withstand the next processing stage (doping, etching, etc.), the resist is thoroughly dried and hardened by baking in an oven. Once the subsequent process has been completed and the resist has served its purpose, it is stripped away.

A1.3.2 Mask alignment

As mentioned earlier, the alignment of each processing layer on top of previous layers is critical to the successful manufacture of the circuit. As the dimensions involved are now of the order of microns or less, a mask alignment process of this accuracy is also required. To aid this process, alignment registration marks are placed on each of the mask layers. These are usually crosses or diamond shapes at each corner of the mask so that the shapes can be fitted within the previous layers' marks. The alignment process can be done manually with the aid of high-power microscopes and split views to aligned marks on two sides of the mask. Computers with pattern recognition systems are being used more for this process, increasing the degree of automation.

There is a choice in the position of the mask relative to the wafer at the point of exposure to light. Firstly, the mask may be held in hard contact with the wafer. In this case the light source does not need to be collimated as there will be no diffraction effects and the pattern is transferred directly. The great disadvantage of this contact process is that damage can occur to the wafer and the mask. Damage to the mask may then be transferred to subsequent wafers. The mask is usually cleaned after about 20 exposures, and this process may be repeated 10 times, but the mask has a very limited life. Alternatively the mask may be held in close proximity, but not touching the wafer. This reduces damage to mask and wafer, but may not eliminate it completely. In this case the light source must be collimated into parallel beams or the edges of the pattern will become blurred. The mask can be used many more times and so it becomes economic to provide a better quality mask in terms of dimensional control and pattern quality. In the contact process the mask is not used very often and so must be as cheap as possible. The third alternative is to keep the mask and wafer a relatively large distance apart. In this case the chance of mutual damage is completely eliminated and the mask can be used thousands of times and more money can be spent ensuring a good-quality mask. The problem with this projection print system is that only a small area of the wafer can be exposed to the light at any one time, as it is impossible to maintain the focus of the source over the whole wafer. Thus the exposure has to be done in a series of 'step and repeat' operations, and is consequently slower than the other two techniques.

A1.3.3 Electron beam lithography

The photolithographic process is based on ultra-violet (u-v) light having a wavelength of the order of half a micron. As the resolution of the process is about the dimension of the wavelength of the light, there are problems in achieving sub-micron resist patterns based on u-v exposure. In order to increase the resolution of the lithographic process, resists that are responsive to electron beams have been developed. These electron beams have a much shorter associated wavelength, of the order of 10^{-6} microns. In practice, the resolution is determined by the quality of the beam forming optics.

The difficulty in using an electron beam is that the process must be done in an evacuated chamber, so throughput is very slow as only a few wafers can be loaded into the chamber at any one time.

An alternative is to use X-ray lithography. X-rays are electromagnetic radiation of the same form as u-v, but of much higher frequency and so shorter wavelength, typically tens of Ångstroms. X-ray lithography is still very much at the research stage and is rarely used in production.

A1.4 Doping and ion implantation

The basic wafer can have a certain amount of 'background' dopant introduced into it at the purification and crystallization stage, and higher doped active layers can be grown on top of the wafer by the epitaxial process described above. However, both of these processes are global, in that the whole wafer is processed. For certain parts of the planar process, it is required that areas are selectively doped in conjunction with the photolithography process. For example, the highly doped source and drain areas of a FET have to be introduced into a background of the opposite type of doping.

As photoresist cannot usually tolerate the high temperatures in which this process is performed, the 'protection' layer must be a dielectric layer (see section A1.5), in which the windows through which the dopant is selectively introduced have been defined by a previous lithography step. There are two basic approaches for introducing the dopant material, diffusion and ion implantation.

A1.4.1 Diffusion

This is the original process by which all selective doping used to be introduced. The basic approach is to expose the area to be doped to a source of the dopant material, and then by elevating the wafer to high temperatures, the material diffuses into the semiconductor. The diffusion process is very well modelled, so for a given temperature, the time of the diffusion to provide a particular depth of penetration or doping profile can be easily calculated. The diffusion can either be done in one stage, or a two-stage process that consists of a lower temperature 'deposition' stage, after which the dopant source is cut off, followed by a higher temperature 'drive-in' stage. The first of these processes is termed continuous source, as a fresh supply of the dopant is continuously available. The second is termed limited source diffusion, and the two

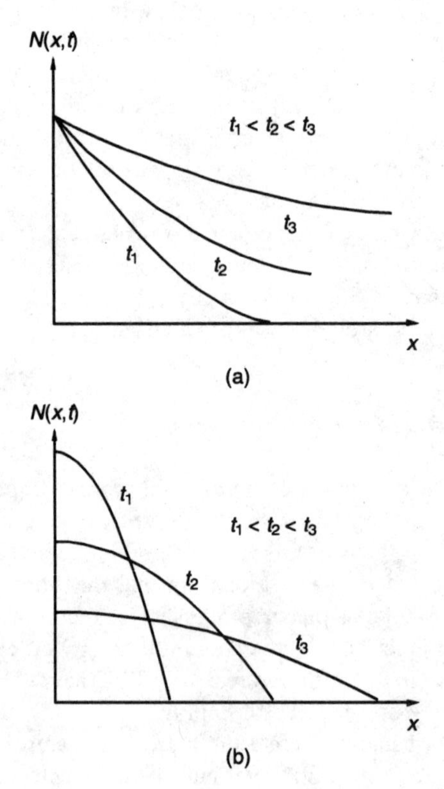

Figure A1.3 Diffusion profiles: (a) continuous source; (b) limited source

processes provide distinctively different profiles with time, as illustrated in Figure A1.3. These two techniques may be used at different stages of the processing, depending on the profile requirements of the various doped areas.

As the diffusion process is accelerated by raising the temperature, and doping profiles have to be carefully calculated for correct device operation, care must be taken with further high-temperature processing of the wafer so that previously processed areas are not disturbed. This is particularly true when there are multiple diffusion processes, for example the formation of base and emitter regions in a bipolar process, as illustrated in Figure A1.4. Here the n-type emitter region needs to be much more highly doped than the p-type base region for efficient transistor action. As the emitter diffusion will be the later diffusion, it is preferable that the temperature of this process is done at a lower

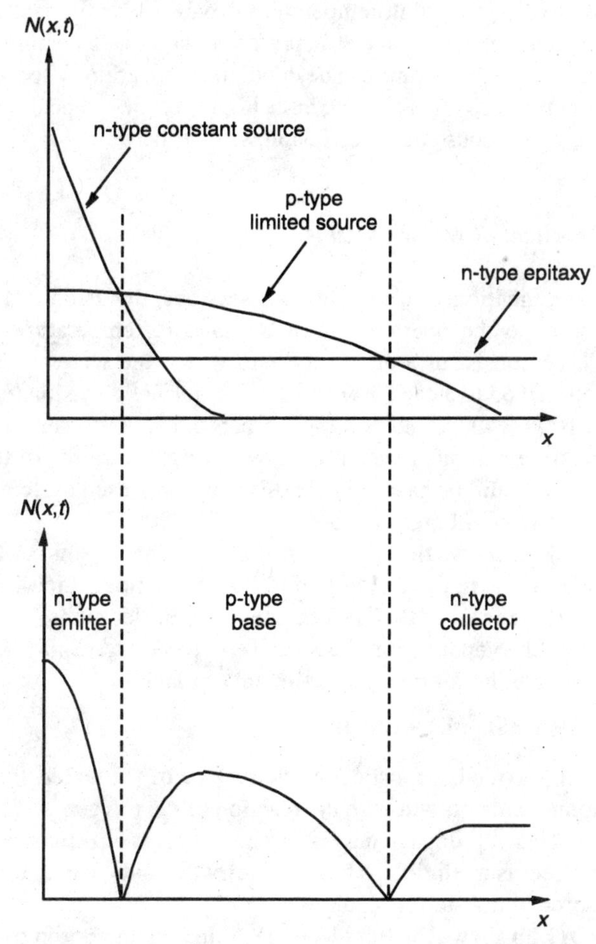

Figure A1.4 Bipolar transistor diffusion profiles

temperature than that of the base diffusion, so that this first diffusion process is not significantly affected by the second process.

There are two considerations here. The n-type dopant (for silicon wafers) is a group V element such as phosphorus or arsenic. The p-type dopant is from group III, such as boron, gallium or aluminium. Each dopant element has a different diffusion constant associated with it which describes how quickly the material diffuses into the wafer at a given temperature. The other consideration is the different profiles available from the continuous or limited source processes. As in this example we would like a relatively low doping density in the base, which will not be affected greatly by the subsequent emitter diffusion, we should choose a dopant with a low diffusion constant, such as gallium, so that a higher diffusion temperature is required. We also require a relatively flat profile in the base, so the limited source approach is preferable with a relatively long, high-temperature drive-in. For the emitter diffusion, a high diffusion constant material is preferred, such as phosphorus, so that this is done at a lower temperature. The depth of penetration is not too great, so the base region is not overly affected, but a high level of n-type dopant is required, so a short, continuous source diffusion will fit the bill.

A1.4.2 Practical diffusion systems

As has been mentioned in the previous sections, diffusion is a high-temperature process, so the operation must be done in temperature-controlled furnaces. These consist of a glass tube into which the wafers are inserted after being mounted on to a carrier or 'boat'. The normal diffusion temperatures are typically 1000–1400°C, so at these temperatures the dopant material must be introduced in a gaseous form, although the original source of the dopant may be of a solid, liquid or gaseous form. Examples of each system are shown for a boron diffusion in Figure A1.5.

Figure A1.5a shows the system for the solid dopant source. In this case the dopant material is in the form of solid boron trioxide which is held in a platinum boat at the front of the furnace tube. The carrier gases (nitrogen and oxygen) pick up the vapour from the heated source and transport it over the wafer surfaces, where the following reaction takes place:

$$2B_2O_3 + 3Si \quad 4B + 3SiO_2$$

The silicon dioxide is deposited on the surface of the wafer. In addition, there will be some oxide growth with the reaction of the oxygen carrier gas (see next section) but as the dopant material readily diffuses through the oxide, any masking effect is negligible. The oxide growth can be easily etched off before the next processing step if required.

Figure A1.5b shows the liquid source system, using boron tribromide which is heated and through which the nitrogen is bubbled, again resulting in a

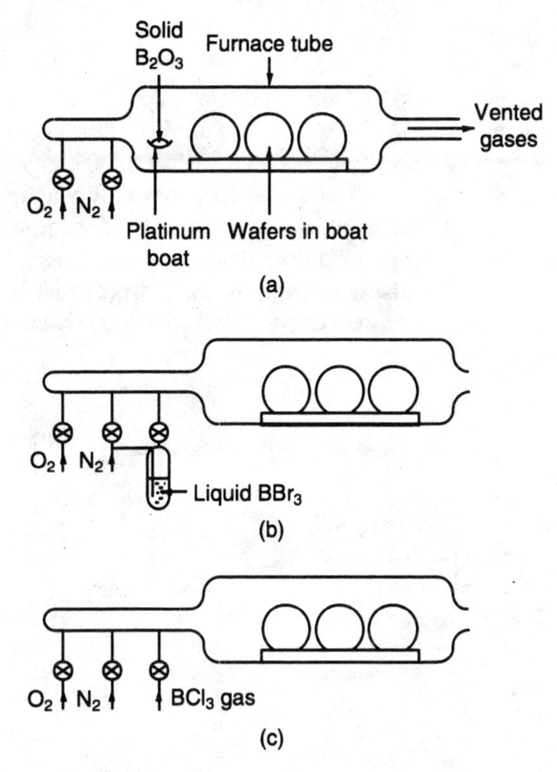

Figure A1.5 Boron diffusion systems: (a) solid diffusion source system; (b) liquid diffusion source system; (c) gaseous diffusion source system

vapour form of the compound. In this case the vapour reacts with the other carrier gas, oxygen, to react accordingly:

$$4BBr_3 + 3O_2 \rightarrow 2B_2O_3 + 6Br_2$$

The boron trioxide, again in a vapour phase, can react with the silicon, as above, to release the elemental boron, while the bromine gas is taken out with the exhaust gas.

Finally, see Figure A1.5c, boron dopant can be introduced to the system in the gaseous form of boron trichloride, again with oxygen as the reaction gas and nitrogen as the carrier gas:

$$4BCl_3 + 3O_2 \rightarrow 2B_2O_3 + 6Cl_2$$

The reaction and process are identical to those for the liquid source, which also involved the boron halide.

A1.4.3 Ion implantation

An alternative approach to doping from the diffusion process consists basically of firing dopant material at the wafer. The high kinetic energy of these atoms means that they can penetrate into the surface of the semiconductor. The faster they are 'shot' at the wafer, the deeper they will penetrate, so control of the doping profile is achieved by the speed of the dopant material.

The mechanism for 'shooting' the dopant material is to accelerate it through an electric field. To do this, the material must be charged, that is consist of ions, so the overall process is termed ion implantation. A typical ion implanter is shown schematically in Figure A1.6.

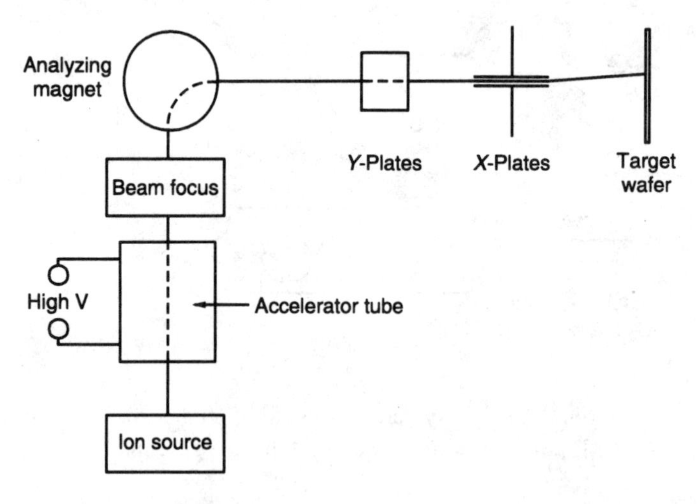

Figure A1.6 Ion implantation system

The ion source consists of the dopant material within a very strong electric field – strong enough to strip the dopant molecules apart into the constituent ions, forming a charged gaseous plasma. These charged ions are then accelerated to the desired velocity by control of the voltage along the accelerator tube, and then focused into a narrow beam of high-velocity ions. The beam is turned through 90° by the use of an analyzing magnet. The purpose of this element is to purify the beam of any unwanted contamination ions which may also have been accelerated with the dopant beam. As any other ions will have a different mass from the desired material, they will be turned by a different amount, the magnetic field having been set up so that only the desired dopant turns through a right angle. The focused beam of dopant ions is then passed through *X*- and *Y*-scanning plates. This mechanism is almost identical to the scanning plates in a cathode ray oscilloscope which move the focused beam over the screen. In this case the ion beam is moved over the target wafer, as the implantation has to be done bit by bit – the whole wafer cannot be implanted in one operation.

As the implanting effect will cause much mechanical damage to the surface region of the wafer, it is necessary to pass the wafer through an annealing process. This consists of raising the wafer to a temperature of around 600°C, and this has two effects. Firstly, the semiconductor wafer can re-align its crystal structure back to the single crystal which is important for efficient device operation; and secondly, it allows the dopant material to fit substitutionally into the crystal lattice.

The great advantage of ion implantation over the diffusion process is that the former is a low-temperature process, so unwanted diffusion of impurities is greatly restricted. However, the ion implantation equipment is much more expensive than a diffusion furnace system, and as the operation has to be performed in an evacuated chamber, the throughput of wafers is very low.

A1.5 Oxide and other dielectric layers

The uses of oxide and other dielectric layers within semiconductor fabrication processes are many and varied, but fall broadly into two categories. Firstly, they can be used as an aid in the actual fabrication process itself, such as a mask against diffusion or ion implantation processes, or as a final passivation layer. Secondly, they can form the integral parts of the final devices and circuits, such as insulation between different circuit layers, isolation between devices, dielectric layers in planar capacitors and part of the gate structure in MOSFETs.

The great advantage in the use of silicon as a semiconductor substrate is the fact that it has a natural oxide, SiO_2, which forms a very good insulating, dielectric layer and which can be easily grown from the native semiconductor by high-temperature exposure to oxygen or water vapour. Alternatively, a non-native form of the oxide can also be deposited over the wafer. This must be done where the oxide is required in areas where the substrate silicon is covered by some other processing layer, such as metallization. An alternative dielectric layer which can also be deposited is that of silicon nitride (Si_3N_4). It is possible but not practical to grow a nitride layer from the native semiconductor. Another deposited layer that is of use in certain circuit processes is polycrystalline silicon. Although not strictly a dielectric layer, it is more like a conductor, the process is similar to others in this section and so is included here.

Each of these layers will now be described in more detail.

A1.5.1 Thermal oxidation

As mentioned above, the native oxide layer may be grown by reaction of the silicon with either oxygen or water vapour, known as dry and wet processes respectively. At the temperatures involved (900° to 1200°C), both oxygen and water vapour readily diffuse through the oxide layer, so the process can continue almost indefinitely and very large thicknesses of oxide can be grown. The two chemical reactions are:

$$Si + O_2 \rightarrow SiO_2 \text{ (dry)}$$
$$Si + 2H_2O \rightarrow SiO_2 + 2H_2 \text{ (wet)}$$

The wet process has a faster rate of growth, but the dry process results in a denser, more stable oxide layer. So, in general, the wet process is used for thicker, insulating and isolation layers, and the dry process is used for thinner oxide layers such as the gate oxide in a FET process, or the dielectric in monolithic capacitors.

A1.5.2 Deposited oxide layers

Where the native oxide cannot be grown because the silicon substrate is masked by other layers, the oxide must be deposited in a CVD process. There are a number of options available for the reactive components, a common one is that of silane and oxygen, the reaction temperature being around 450°C:

$$SiH_4 + 2O_2 \rightarrow SiO_2 + 2H_2O$$

Sometimes the deposited oxide layer is doped with phosphorus. The advantage of this is that, for certain doping levels (6–8 per cent P by weight), the resulting layer can be made to soften and flow, providing good step coverage and a more even surface for subsequent processing steps.

A1.5.3 Nitride layers

Nitride layers find use as an oxidation mask in certain MOS technologies, gate dielectrics in EEPROMs, high permittivity layers for capacitor structures and also as a final passivation layer as a good barrier against moisture and contamination by sodium ions, which diffuse rapidly through oxide layers.

There are two main CVD techniques for the deposition of nitride layers, the reaction of ammonia with either silane or dichlorosilane:

$$3SiH_4 + 4NH_3 \rightarrow Si_3N_4 + 12H_2$$

at 700° to 900°C and atmospheric pressure

$$3SiCl_2H_2 + 4NH_3 \rightarrow Si_3N_4 + 6HCl + 6H_2$$

at 700° to 800°C and reduced pressure.

A1.5.4 Polycrystalline silicon

Although not strictly a dielectric film, the deposition of polycrystalline silicon (or polysilicon for short) is included here as the techniques are similar to the other films described above. Polysilicon differs from the normal silicon structure that forms the substrate and epitaxial layers in that the material is not formed of a single crystal but consists of small grains of single crystals at random orientations to each other. Like single-crystal silicon its conductivity can be altered by doping. The relationship between doping density and conductivity is not a straightforward one, as some of the dopant material can reside at the grain boundaries and be inactive.

Polysilicon films are used where it is not possible to grow an epitaxial layer because the substrate is covered by another film (oxide or metallization, for example). It finds use in interconnect lines in multi-level conductor systems, in which case the polysilicon is highly doped, and in passivation systems, in which it is undoped or even sometimes has added oxygen to increase the material resistivity.

Polysilicon is formed in a low-pressure furnace at 600° to 650°C by pyrolyzing silane:

$$SiH_4 \rightarrow Si + 2H_2$$

A1.6 Etching

A major part of the fabrication process is the need to (usually selectively) remove the various layers. The generic process here is called etching, and can be considered in two forms. When IC fabrication was first developed, and for many years afterwards, the etching process was achieved by immersing the wafers in a liquid chemical etch. The particular chemical etch depended on the layer being removed (silicon, silicon dioxide, metal, photo-resist, etc.). If a selective etch was being performed, the areas on the wafer that were not to be etched were protected by a resist or possibly oxide layer whose pattern had been defined by a photolithography stage. This form of etching is known as 'wet' etching.

The main problem with a wet etch is that of undercutting below the protective layer, as illustrated in Figure A1.7. As the etching progresses, the liquid etch is free to swirl about and so generally will etch in an isotropic way, that is at the same rate in all directions. This means that the amount of undercut-

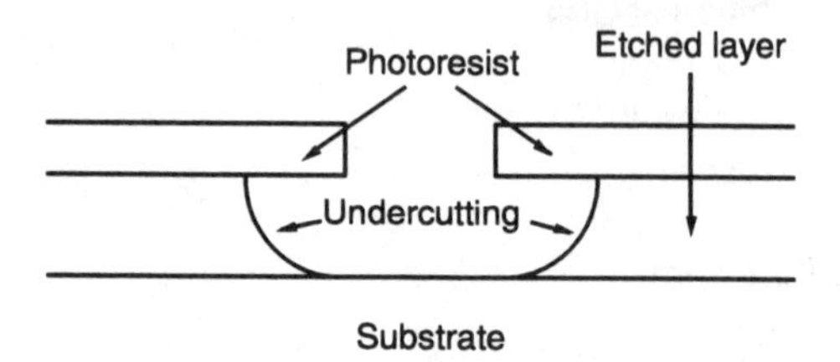

Figure A1.7 Undercutting effect of wet etch

ting, and hence reduction in feature size from that specified by the mask geometry, are about the same size as the layer thickness being etched. Provided that the feature sizes are much larger than the layer thicknesses, this presents no real problem. In later years, however, as the feature sizes on ICs have progressively reduced, it became possible for undercutting to have a serious tolerancing effect, to the extreme case of features or lines being completely removed.

A solution needed to be found in terms of obtaining an anisotropic form of etch in which the etch rate is much faster in the vertical direction than the horizontal direction. There are a number of forms of this etching, such as plasma etching and reactive ion beam (RIB) etching, and these are grouped under the general heading of 'dry' etches. The wet etch is a much cheaper and simpler approach in general than the dry etch, but for feature sizes of the order of a micron, a dry etch is almost essential and has largely supplanted wet etching.

A1.6.1 Wet etching processes

The actual chemicals and conditions for the etch depend on the layer being etched. The usual layers etched with a wet etch are silicon (generally in the wafer preparation stage to provide a mirror-smooth finish), silicon dioxide (generally for definition of doping areas), metallization (often aluminium to define the conductor interconnects) and photoresist (complete stripping of layers that have served their purpose).

Because of the crystal structure of the silicon wafer, it is possible for a wet etch to act anisotropically on silicon, etching at different rates along different crystal orientations. There are several variations of etch for silicon, most being based around a combination of hydrofluoric (HF) and nitric (HNO_3) acids.

Perhaps the most common etch procedure is that of the removal of silicon dioxide, as this can occur at several different fabrication stages, including definition of diffusion areas, contact holes between conducting layers and opening bonding windows in passivation layers. The principal oxide etch is based around HF in a buffered solution. This will etch silicon dioxide much more rapidly than silicon or photoresist. The etch rate is also dependent on the temperature, which has to be carefully controlled, whether the oxide was grown wet or dry, and also any dopant material within the oxide layer. As the etch will also remove the silicon below the oxide being etched, albeit slowly, end point detection is very important. Fortunately there is a characteristic of silicon dioxide that makes visual end point detection very easy. The characteristic is that water and HF will 'wet' silicon dioxide, that is the liquid will stick to the layer, whereas they do not 'wet' silicon, and the liquid easily runs off this material, leaving it virtually dry. This change is clearly visible in the surface of the wafer, which can be withdrawn from the etch immediately after 'de-wetting' has taken place.

Etching of photoresist is very straightforward as these layers are based on organic materials, so an organic solvent such as acetone will easily strip the resist layer while leaving the other layers, based on inorganic materials, untouched.

The etching of metallization layers is also relatively straightforward. The most common metal used for these layers is aluminium, although gold is used in some applications, for example with GaAs substrates. The standard process of depositing a film of metal, then defining a layer of photoresist and subsequent etching is one approach that can be adopted for defining the metal interconnects. Because of its nature, this is termed a subtractive process. However, there is an alternative approach termed lift-off. In this, the first layer is a resist layer which is then patterned in the reverse sense to that used in the subtractive approach. Once the pattern has been defined in the resist layer, the metal is deposited over the wafer; the wafer is then immersed in an organic solvent which removes the resist layer. As this layer is dissolved, the metal over the resist is lifted off into the solution, leaving behind the desired metal pattern.

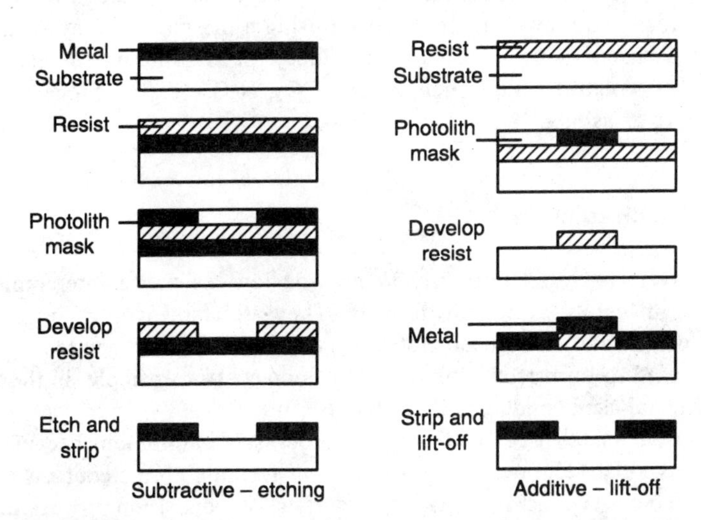

Figure A1.8 The two forms of metallization lithography

This lift-off process is referred to as an additive process. Both the additive and subtractive processes are illustrated in Figure A1.8.

A1.6.2 Dry etching

The problems of the isotropic wet etch were highlighted earlier. To overcome these problems and to achieve finer dimensional resolution, a number of techniques for dry etching of various layers have been developed. There are three basic approaches to dry etching which are described below.

Plasma etching involves a vacuum system from which the atmosphere is removed. Then a small amount of a reactant gas, such as chlorine, is introduced. This is excited by a high-strength radio frequency field to form a plasma of charged ions. These then react with the silicon dioxide to form volatile silicon compounds which evaporate off from the wafer.

Sputter etching also employs ions, usually from heavy noble gases such as argon. These are not reactive but instead are accelerated through an electric field and 'fired' at the wafer. The process is one of simply using the kinetic energy of these heavy ions to knock material off from the topmost layer of the wafer. Very anisotropic etching can be achieved, but selectivity is very poor, so metal layers often have to be used as a barrier to wafer etching.

Reactive ion etching (RIE) or reactive ion beam etching (RIBE) is a combination of the two previous processes in that ions are once again formed using a plasma system, and these are then fired at the target wafer. The etching occurs by a combination of reaction of the ions with the etched layer, and energy transfer of the accelerated ions. With the advantages of both of the techniques, good control can be obtained of the etching process.

Photoresist stripping can be achieved using a plasma-based system, where oxygen is used as the reactive ion source, the organic-based resist being reduced to volatile oxides such as CO, CO_2, H_2O etc. The process is often referred to as ashing.

A1.7 Metallization

Metal layers are used primarily to provide low-resistance interconnections between different devices on the chip. They are also involved in providing ohmic contacts to silicon substrates or polysilicon interconnections and may also be used to form rectifying Schottky contacts (for example, in the formation of metal–semiconductor FETs (MESFETs)).

The usual metal used in the silicon process is aluminium, because of its cheapness, relatively low resistivity, ease of forming ohmic contacts and low melting point. The last point is important, as the deposition process of metal films is usually that of evaporation. A typical evaporation system is illustrat-

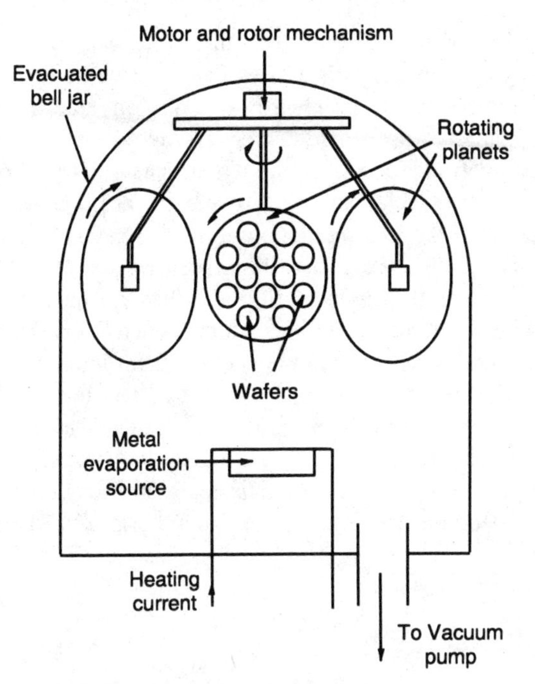

Figure A1.9 Metal deposition system

ed in Figure A1.9. Many tens of wafers can be accommodated at a time. These are loaded into 'planets' which rotate or 'orbit' around the metal source during the deposition process to achieve an even rate of deposition. Once the wafers are loaded the system is evacuated before the evaporation process takes place. There are two basic approaches to evaporation. The simpler approach is to load a boat made of a high melting point metal, such as tungsten, with the aluminium. A large electric current is passed through the boat, heating it to a temperature above the melting point of aluminium. Being in a vacuum the metal then evaporates, is spread through the system and deposits in a solid form on the unheated wafers. The alternative approach to evaporating the metal is to fire an electron beam at the source of metal to vaporize it.

The rate of deposition of evaporated metal is relatively slow, but for most applications where only relatively thin films are required (of the order of a few microns) the process is satisfactory. For some specialized applications (such as forming heatsinks for high-power circuits or for high-frequency circuits where skin depth effects are a consideration), thicker metal layers are required than are practical from evaporation systems. Two other approaches are available. The first is based on a sputtering system, similar to the etching system described above; here high-energy noble gas ions are fired at a target of the metal to be deposited, and these evaporate the metal which is then deposited on the target wafer. The disadvantage of this system is the slow throughput.

The alternative is a plating system; here the wafer must have a thin layer of the metal already evaporated on it. It then forms the cathode of an electrolysis system and is immersed into an electrolyte solution containing a salt of the metal to be deposited. An electric current is passed through the system and further metal is electrodeposited on the wafer. The quality of plated films, in terms of the film purity, thickness control and surface topography is relatively poor, but plating is the only realistic option for very thick films.

For some applications, refractory (high melting point) metals need to be deposited. This may be for the formation of Schottky contacts or in combination with silicon to form metal silicides which have a lower resistivity than the deposited polysilicon layers, and are hence better for interconnections. Owing to the high melting point, an evaporation process is not applicable, and so these metals (such as tungsten, molybdenum and palladium) are deposited using a CVD process similar to that of the epitaxial system described earlier. The metal is typically introduced into the reaction furnace as a halide which may be directly decomposed into an elemental form, or reduced by reaction with hydrogen.

A1.8 Passivation and assembly

Once all the processing stages have been performed to create the electrical active circuit, the separate chips have to be mounted into suitable packages that can be incorporated into board-level systems.

The first stage of this process is to add a layer of protection to the circuits. The chip has been fabricated under clean room conditions where the chances of contamination from dust, water, or any other unwanted materials is reduced to a minimum. Before the chip is released to the outside world, as much of the surface as possible must be covered by a protective layer to minimize any subsequent contamination. This is generally achieved by the deposition of a further layer of silicon dioxide or nitride across the whole surface, so providing a reasonable protection against dust and waterborne contamination in particular. However, there still have to be electrical connections made to the input and output points on the chip, so a further photolithographic process has to be performed to open windows in this passivation layer and allow the next processing stage to take place.

At this point there are typically hundreds or thousands of chips on a single wafer which will have to be individually separated and packaged. It is possible to do electrical tests at this time to root out defective chips and so avoid the time and cost of packaging devices that will subsequently be found to be defective. This is done with a probing system where metal needles are brought into contact with the input/output (I/O) pads. Defective chips are identified by placing a drop of ink on them after the prober has tested them. The prober operates in a 'step and repeat' mode, covering the whole wafer in a programmed fashion.

The next task is to separate the individual chips. The most common technique is one of 'scribe and break', whereby a diamond-pointed scriber is run down the horizontal and vertical scribe channels between the individual chips. If the wafer is then bent, it should preferentially break along a scribed groove. To ensure all the chips separate, the wafer is mounted on a sticky rubber backing which is then stretched. The breaks occur along the scribe lines and the individual chips are held to the rubber layer for subsequent processing.

The chip is next mounted in a package that normally consists of a frame with a number of metal leads connected to 'legs' which will form the connections that can be soldered on to the circuit board. The chip has to be held in place in the centre of this frame, which is usually achieved by means of a spot of epoxy glue. In some cases the base of the substrate also forms an electrical contact, in which case the chip has to be mounted using a conducting epoxy or low-temperature solder. The electrical connections that have to be made from the I/O pads to the frame leads are achieved by bonding very thin metal wires from the bond pads to the leads. The mechanical contact may be accomplished by applying an ultrasonic signal at the point on bonding ('scrubbing' the bond wire on to the contact) or by a thermo-compression technique which is a combination of local heating and high pressure when making the bond. Again the bonding is often achieved in a step-and-repeat procedure which is programmed into the automatic bonding equipment.

Once the electrical contacts are complete, the chip is hermetically sealed into the package to provide a further layer of protection from contaminants that may cause premature failure of the device. The usual forms of packaging are a plastic compound that is moulded around the bonded chip, or a ceramic-based package which is more expensive but which provides better stability and greater thermal range.

A1.9 Examples of basic processes

To complete this appendix, examples are given of the three most important circuit processes: bipolar, nMOS and CMOS. The processing steps and mask levels associated with each are highlighted. These descriptions provide important background material to the techniques described in the main text, in particular the transistor-level design techniques in Chapter 3. Each of the three descriptions consists of a series of drawings of the cross-section of the wafer at each particular stage with a brief description of the process involved at that stage. It should be borne in mind that the fabrication process for each example is the simplest possible to realize a device, and that the drawings are simplified and not to scale.

A1.9.1 Bipolar process

The bipolar process is illustrated in Figure A1.10a–i. The circuit illustrated here simply consists of a single n–p–n transistor, with bonding only to the collector contact. Starting with a p-type doped substrate (a), the first pho-

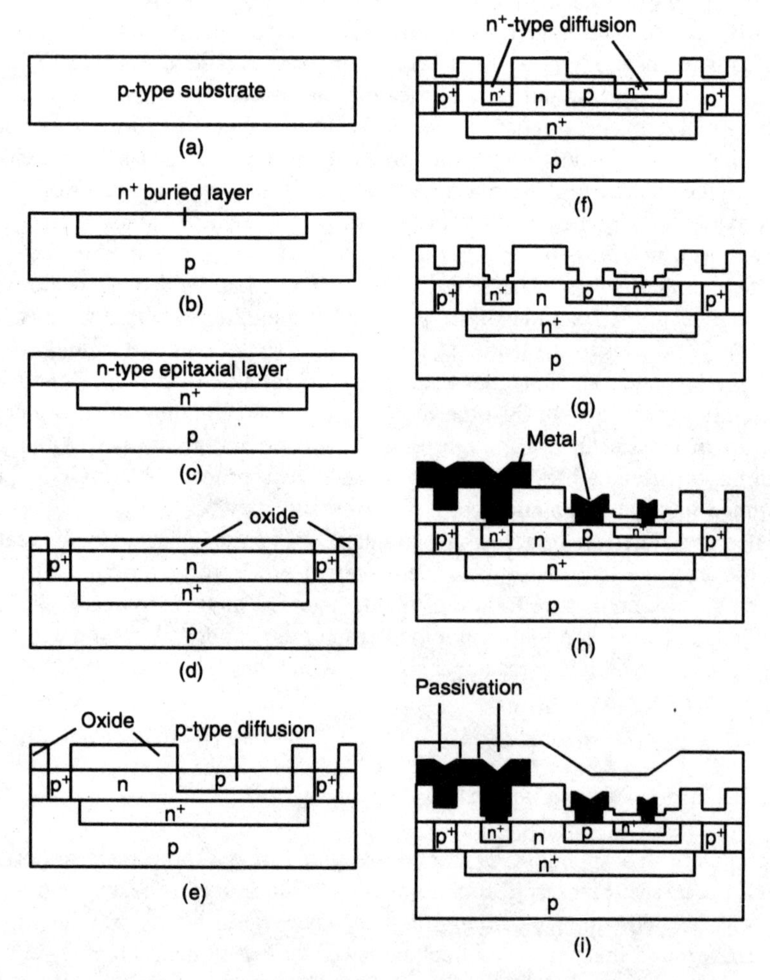

Figure A1.10 Bipolar fabrication process: (a) initial substrate; (b) buried layer diffusion (mask 1); (c) epitaxial layer growth; (d) isolation diffusion (mask 2); (e) p-type base diffusion (mask 3); (f) n-type emitter diffusion (mask 4); (g) metal contact openings (mask 5); (h) metallization (mask 6); (i) passivation and bonding cuts (mask 7)

tolithography stage defines where the buried n-type layer will go (b). This highly doped region provides a low-resistance path for the collector. Next, an n-type epitaxial layer is grown over the whole wafer (c). Highly p-doped regions are defined in the next mask process, with subsequent diffusion to form channel stops, or insulating regions between devices (d). The next stage is to define the base areas and dope them p-type (e). Next the highly doped emitter regions are defined and doped (f). Contact cuts are made in the insulating oxide layer so that the metallization can form ohmic contacts to the emitter, base and collector regions (g). The metal is deposited and the interconnection pattern formed in the next mask layer (h). Finally the passivation layer is deposited, and holes through which the bond wires will contact are defined (i).

A1.9.2 nMOS process

The nMOS process is illustrated in Figure A1.11a–i. The circuit consists of two transistors, one enhancement mode and one depletion mode, as may be configured in a simple digital inverter (see Chapter 3). Being based entirely on n-channel FETs, the base substrate must be p-type (a). The first stage is to isolate the various active areas where the transistors are to be formed, which is done by depositing a nitride layer and patterning this to protect the active regions. This acts as a barrier for a field implantation of a high concentration of p-type dopants and also the growth of a thick oxide (b). The nitride has then served its purpose and is removed. For the depletion mode device, a thin n-channel must be implanted, and this is defined by mask 2 (c). A thin gate oxide is then grown over the active region (d). The gate structures themselves are formed in polysilicon, as it is not possible to grow an epitaxial layer above the gate oxide. The gate structures of polysilicon and oxide are defined by mask 3 (e). These gate structures can be used as a protecting layer for the next stage which is the diffusion of the n-type regions that form the source and drain regions (f). Such a technique is termed 'self-aligned gate'. A further layer of insulating oxide is deposited to protect the gates from the subsequent metallization, and contact cuts made where electrical connections are required between the layers (g). The final stages are similar to the bipolar process. Metal is deposited and patterned (h), and the passivation layer is deposited and bonding holes cut (i).

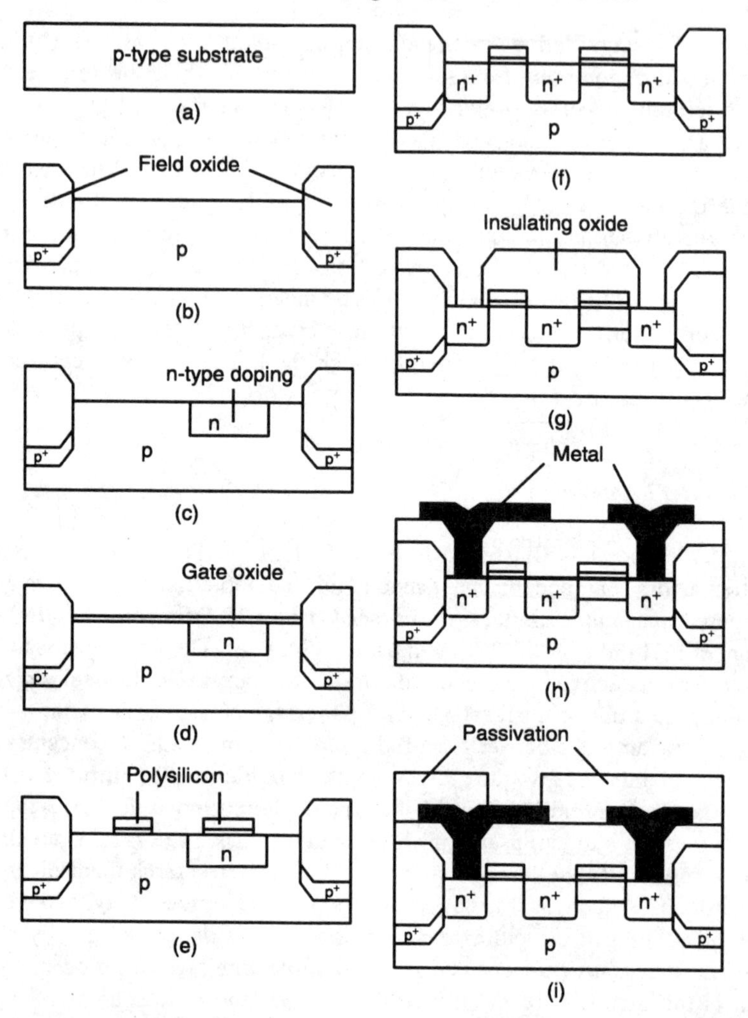

Figure A1.11 nMOS fabrication process: (a) initial substrate; (b) field implant and oxide (mask 1); (c) depletion region implant (mask 2); (d) growth of thin gate oxide; (e) definition of gate structures (mask 3); (f) source and drain diffusions; (g) contact cut definition (mask 4); (h) metallization (mask 5); (i) passivation and bonding cuts (mask 6)

A1.9.3 CMOS process

The CMOS process is illustrated in Figure A1.12a–k. The circuit again consists of two transistors, one n-channel and one p-channel configured as a digital inverter (see Chapter 3). As both types of FET are present, we have the

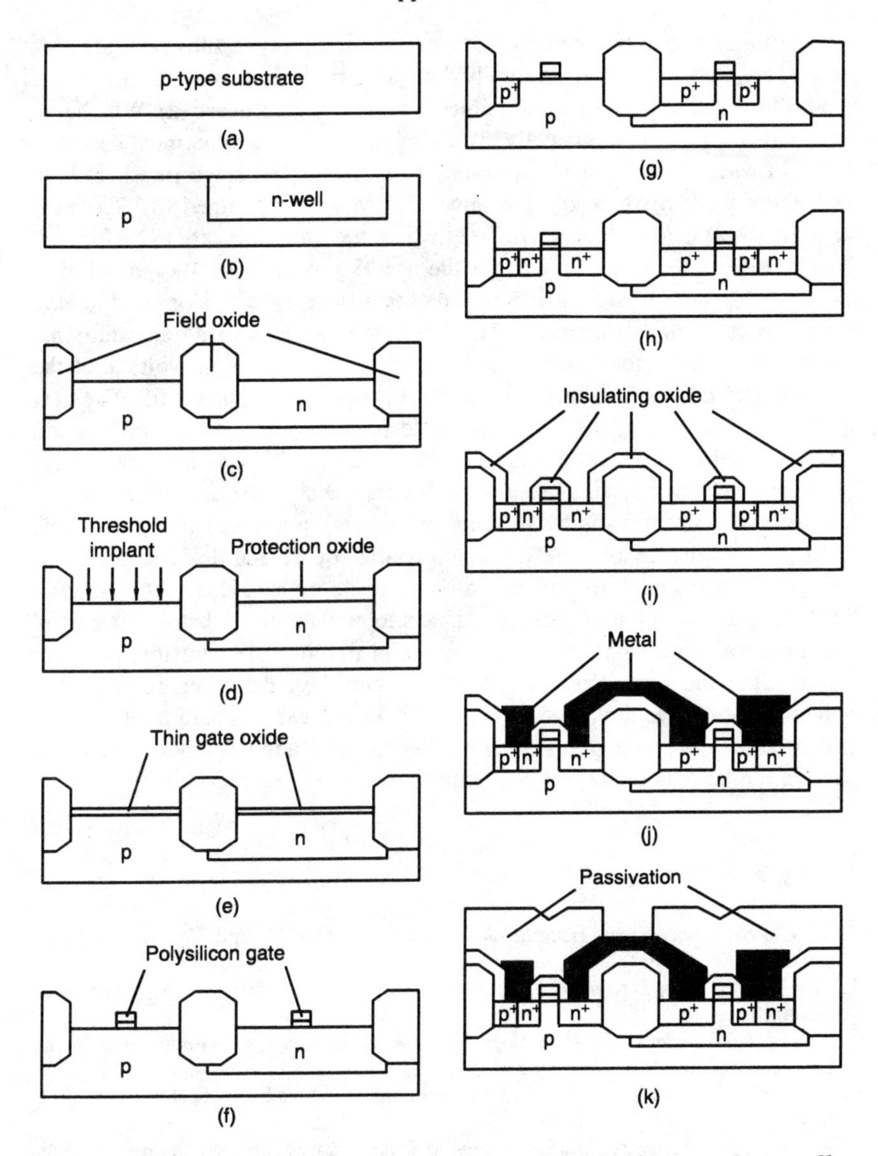

Figure A1.12 CMOS fabrication process: (a) initial substrate; (b) n-well definition (mask 1); (c) field oxide growth (mask 2); (d) threshold implant (mask 3); (e) gate oxide growth; (f) gate definition (mask 4); (g) p^+ diffusions (mask 5); (h) n^+ diffusions (mask 5 reversed); (i) contact cut definition (mask 6); (j) metallization (mask 7); (k) passivation and bonding cuts (mask 8)

choice of either an n-type or p-type doped substrate; n-type substrates are easier to form, and were used originally, but more recently p-type substrates are more often used as they are compatible with the nMOS process (a). Whichever substrate is used, areas, or wells, of the opposite doping type must be introduced in which to form the other type of FET. The first stage (mask 1) is to form the n-well in which the p-channel device will be formed (b). The next stage (mask 2) is to isolate active areas where the transistors are to be formed. This is done in the same way as for the nMOS process with the growth of a thick field oxide (c). As it is difficult to obtain the correct background doping for a particular transistor operation, it is often required to do an additional implant in the p-regions to control in particular the threshold voltage of the resulting transistors. This threshold implant is defined by mask 3 (d). The next stage is the growth of the thin gate oxide (e), deposition of the polysilicon gates, and subsequent patterning defined by mask 4 (f). The next two stages can be defined by the same mask pattern, either using the same mask and different resist polarities, or by forming the inverse mask and using the same resist. The stages involve the formation of the source and drain regions – p-type for the p-channel devices and n-type for the n-channel devices. In addition, these diffusions will provide contacts to connect the substrate to ground and the n-well to the positive supply, in order to eliminate parasitic transistor action. First the p-regions are formed (g) and then the n-regions (h). The remaining processes are identical to the nMOS process: deposition of an insulating oxide with contact cuts (i); metal is deposited and patterned (j); and the passivation layer is deposited and bonding holes cut (k).

Bibliography

J. Allison, *Electronic Engineering Semiconductors and Devices*, 2nd edn, McGraw-Hill, London, 1990.

L.E.M. Brackenbury, *Design of VLSI Systems – A Practical Introduction*, Macmillan, Basingstoke, 1987.

R.L. Geiger, P.E. Allen and N.R. Strader, *VLSI Design Techniques for Analog and Digital Circuits*, McGraw-Hill, New York, 1990.

R.C. Jaeger, *Introduction to Microelectronic Fabrication*, Addison-Wesley, New York, 1988.

D.V. Morgan and K. Board, *An Introduction to Semiconductor Microtechnology*, 2nd edn, Wiley, Chichester, 1990.

S.M. Sze, *VLSI Technology*, McGraw-Hill, Singapore, 1983.

Appendix 2: IC Design Example

A2.1 Introduction

This appendix describes an example of a simple IC design and realization. This forms a design exercise completed by final students on a BEng degree course in Electronics and Communication Engineering at the University of Bath. It illustrates the ease with which a real IC can be successfully designed, manufactured and tested by people with virtually no previous experience.

The various stages are described in the same sort of order as the structure of this book, acting as an example for the steps taken in realizing an IC. There are a number of constraints put upon the design and manufacture, which will become clear at each stage.

A2.2 Design specification

The circuit to be designed and built is based on a linear shift register (LSR) system. This is a digital logic system employing shift registers and exclusive-OR components. The operation of such systems can be described in terms of linear equations, which give rise to characteristic polynomial solutions and can be used for their design.

One application of LSRs is to generate pseudo-random binary sequences (PRBS). These sequences have apparently random properties and so can be used to generate white noise in analogue test systems or to generate a random digital data sequence for transmission system testing. LSRs can also be used to scramble an information sequence, using the random signal generation property, thus making the transmitted signal appear to have random properties. The information is recovered at the receiving end by using an equivalent descrambling circuit, also based on the LSR.

A2.2.1 Theory

The outline of the theory is illustrated in Figure A2.1. Figure A2.1a defines the delay operator D convention. A binary word A is loaded into an n-bit register and then shifted out. D represents unit delay. Hence:

$A = [a(0), a(1) \ldots \quad a(n-1)]$
$A(D) = a(0) + a(1).D + a(2).D^2 + \ldots \quad a(n-1).D^{n-1}$

A typical feedforward shift register configuration in shown in Figure A2.1b, where the ⊕ symbol represents the EX-OR function. This circuit has the transfer function $G'(D) = Y(D)/X(D)$. For this particular case

$$G' = D + D^2 + D^5$$

The more usual form of operational feedforward register is as shown in Figure A2.1c, giving the transfer function $G(D)$:

$$G(D) = 1 + G'$$

Using this feedforward register to form a feedback system as in Figure A2.1d, we have a new transfer function:

$$\frac{Y(D)}{X(D)} = \frac{1}{G(D)}$$

The transfer functions of the feedforward and feedback systems are therefore complementary and can be used as a scrambler/descrambler pair.

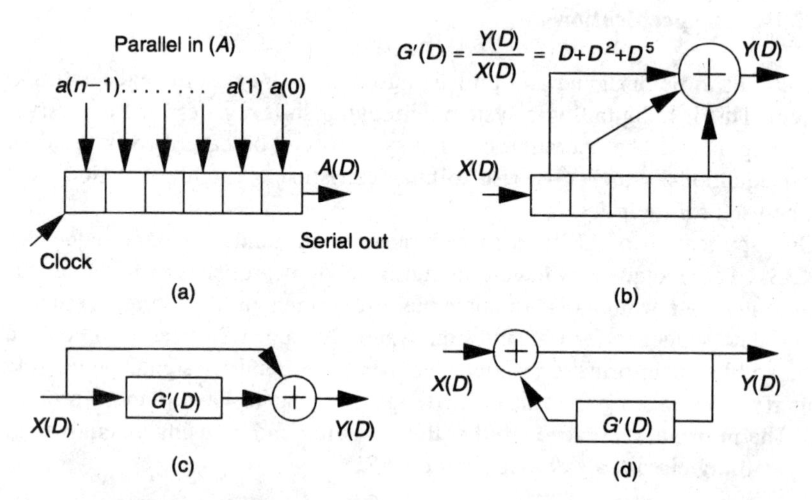

Figure A2.1 Linear shift register circuit principles: (a) shift register conventions; (b) shift register example; (c) feedforward register; (d) feedback register

A2.2.2 Circuit specification

The IC is to contain three identical LSR circuits (A, B and C). Each one consists of 15 stages ($n = 15$) and can have four possible feedforward polynomials. Selection of the active polynomial is by two logic control inputs, SP1, SP2, according to the specification given in Table A2.1.

Table A2.1 *Polynomial definitions*

Control SP1, SP2	*Polynomial G′(D)*	*Register taps*
0 0	$D^3 + D^7$	3, 7
1 0	$D^5 + D^9$	5, 9
0 1	$D^2 + D^{11}$	2, 11
1 1	$D^{14} + D^{15}$	14, 15

The circuit must generate a State Output signal (ST) which is to be high when the register contains the word 10000... (the 1 being at the left-hand, input, end of the register). This output is to be clocked and latched, and can be used as a synchronizing or triggering signal.

Each register can also be set into one of two modes, feedforward or feedback, by the configuration control signal, CC. With CC = 0 the register is in feedforward mode, and with CC = 1 it is in feedback mode, corresponding to Figures A2.1c and A2.1d respectively. Each of the three circuits also has a sequence input, SI, and sequence output, SO. In addition there is a clock input and a general reset input. Also included is a reset control input just for register C.

A2.3 Design and manufacture technologies

The manufacturing route available for this exercise was limited to that available through the UK ECAD and EC Eurochip Initiatives, the University of Bath being a member of both. The design software used was the SOLO 1400 suite of programs developed by European Silicon Structures (ES2), the ICs being manufactured by the ES2 foundry. The technology is standard cell, based on 1.5 micron CMOS. The SOLO 1400 package contains all the programs and libraries necessary for complete realization of the ASIC (schematic capture, simulation, layout, etc.). The same design has also been realized using an erasable PAL technology, but the architecture of these circuits makes them unsuitable for this particular circuit.

A2.4 Design approach

The design of the circuit is relatively straightforward, although there are a number of points that the students had to bear in mind while producing the design. The circuit was broken down into a number of hierarchical levels, the

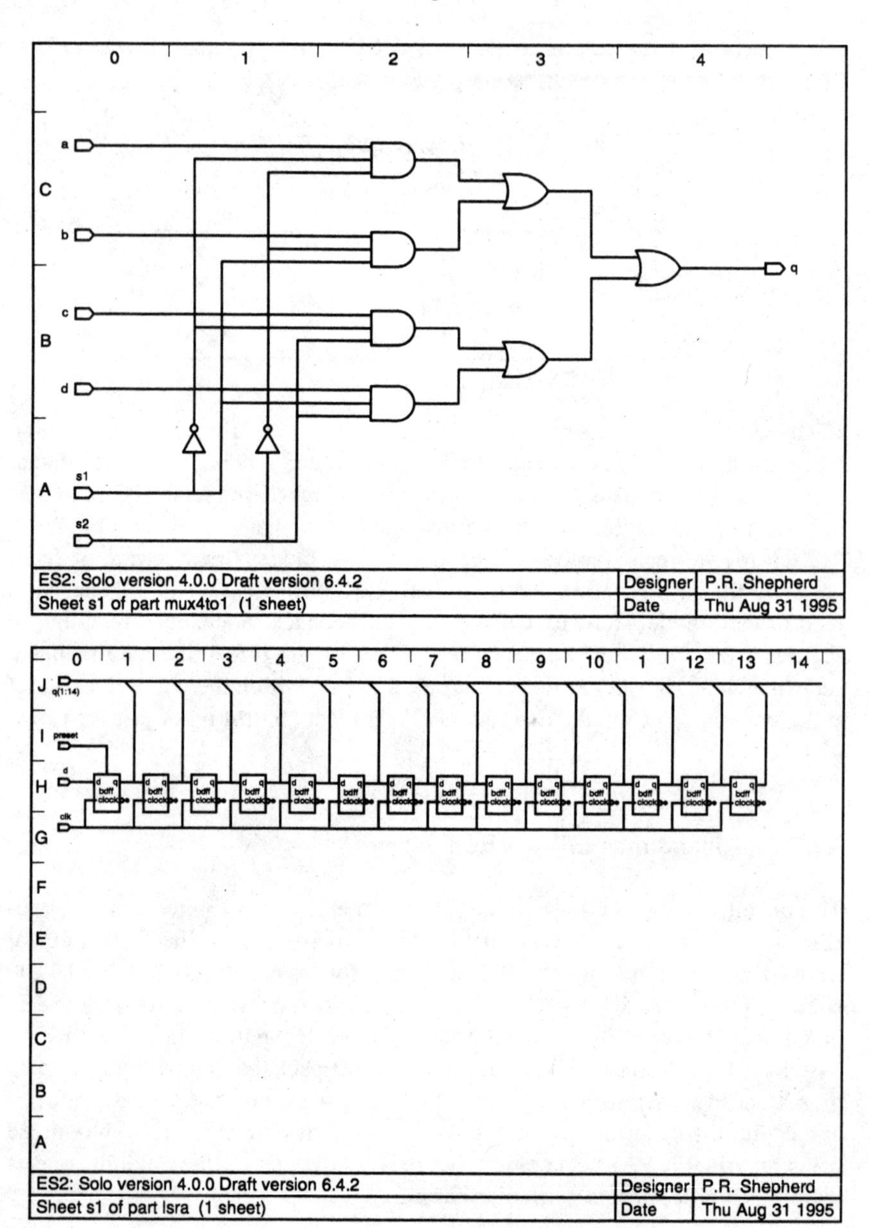

Figure A2.2 LSR circuit design

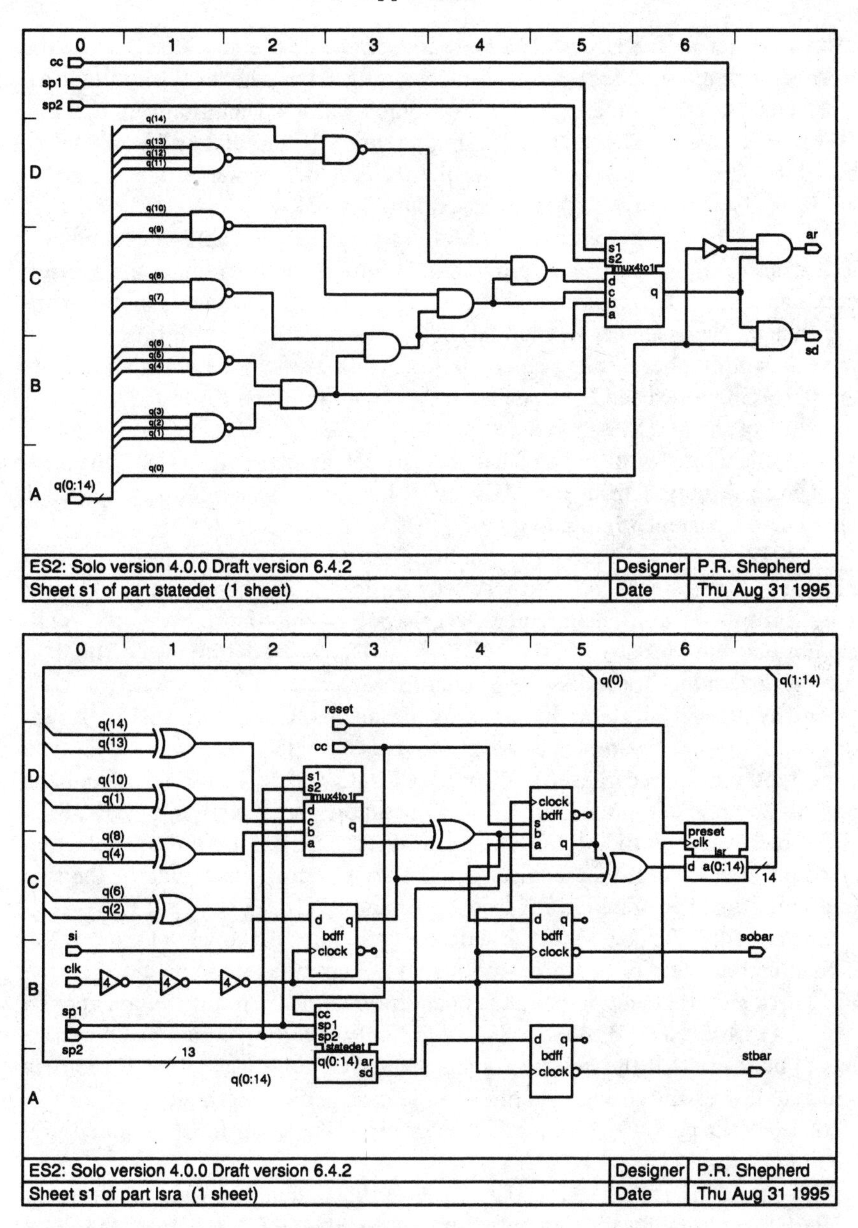

Figure A2.2 *(continued)*

lowest (as far as the circuit design is concerned) is the gate level, using the library elements supplied (no custom designing at the gate level was allowed).

The next level of functional block designed was a 4:1 multiplexer. This circuit is used in the polynomial selection, having SP1 and SP2 as the two selection bits, the four data inputs being the EX-OR'd polynomial taps. The 4:1 MUX is also used in the state detect circuit (see below).

The next level of hierarchy includes the state detect and auto-reset modules. The state detect circuit must provide a 1 output (latched) when the register contains 1000...0. One minor difficulty here is that the width of the word depends on the selected polynomial, being either 7, 9, 11 or 15 bits. NORing the relevant number of bits (except the first) will produce a 1 when all the bits are 0. Having a cascaded tree such that the result of bits 2–7 is input to the first data line of the MUX, 2–9 to the second data line, etc., and using SP1 and SP2 as the select bits, will ensure that the correct number of bits are checked. ANDing the output from the MUX with bit 1 will generate the required 1 which can be latched into a D type.

In PRBS circuits the all-zeros condition must be avoided or the circuit will remain in that state. So an auto-reset circuit detects this condition and feeds a 1 back if this occurs. The circuit is largely common to the state detect circuit, except that the output from the MUX is ANDed with the inverse of bit 1 to provide a 1 output for the all-zeros condition.

The next level of hierarchy contains the main LSR circuit. This is in turn divided into three main blocks. The main register itself consists of 13 D-type flip–flops configured as serial-in, parallel-out. The feedback taps from the appropriate register outputs are EX-ORed and the signal, chosen by SP1, SP2, is fed back through the MUX. The third block consists of the mode selection circuit which is under the control of signal CC. This selects one of the two configurations illustrated in Figures A2.1c and A2.1d.

The core circuitry based on these design principles is shown in Figure A2.2. This illustrates one of the three duplicate LSR circuits that is included on the IC. There are input and output pads in addition to these circuits. In the version of SOLO 1400 used (V3.1), the library has only inverting output pads, so care has to be taken that the inverse signals are fed to these pads. The IC requires separate power and ground supplies to the core and the periphery, so there is a total of four pins – two V_{DD} and two ground. The IC is to be mounted in a 28 pin DIL package.

The circuit is drawn using the SOLO 1400 Schematic Capture program (DRAFT) and the library elements from the ECPD15 (1.5 micron) technology (this has now been superseded by the ECPD10, 1 micron process). From the circuit drawing, a netlist version of the data must be derived for use by the simulator. This is created by first converting the drawing into a HDL form. SOLO 1400 uses its own HDL (MODEL),where the file can be automatically generated before exiting DRAFT. The HDL is a purely text version, so it is possible to view or edit the file before simulation. (It is of course possible to create the design purely in HDL, but this should only be attempted by experienced designers.) The MODEL code is compiled into a netlist in intermediate design language (IDL). This IDL file will be used by the subsequent simulation and layout programs.

A2.5 Circuit simulation

Before progressing to the layout stage, it is vital to ensure that the design is functioning as required to fulfil the specifications, and that consideration is given to the testing procedures to be applied. The circuit is relatively small and there is little likelihood of deeply buried logic, so controllability and observability are high. As detailed in Chapter 6, testing can be functional (checking that the circuit works as expected) or systematic (checking for manufacturing defects). Within this realization route, the functional testing (at simulation stages and after delivery of ICs) is done by the customer. Systematic testing is performed by ES2. One of the requirements for an acceptable design is for the customer to provide a set of test vectors that has a guaranteed node toggle count in excess of 95 per cent.

At this stage, however, we are more concerned with functionality. Having adopted a hierarchical approach, each block (for example, MUX) can be simulated individually to ensure correct functionality. The functionality of the whole circuit must also be checked. The LSR is a somewhat complicated circuit to verify, owing to the pseudo-randomness of its operation. The correct operation of the circuit is (as with all digital circuits) known or predictable. The students wrote a computer program to generate the expected output from a set of inputs, based on ideal operation of the circuit as designed. The results from this program were checked against the output from the SOLO 1400 simulator package (MADS). Sixteen separate tests were devised to check the functionality of the completed circuit. In addition, a test vector file was constructed for submission with the design.

Few functionality problems were uncovered at this stage, it was found that the clock signal had to be buffered as it was used to drive a total of 54 flip–flops. An estimate was made of the maximum speed of the circuit, first by considering the largest signal delay, based on gate delay information in the libraries. The maximum speed was also determined during the simulation by increasing the clock speed until errors occurred. The figures at this stage are not really significant as this is before the circuit is laid out, and parasitics and line delays will have a large effect.

A2.6 Circuit layout

The layout stage of the SOLO 1400 suite, consisting of placement and routing, is entirely automatic, with no intervention from the designer except for specifying certain parameters and the pinout pattern. The structure of the standard cell IC consists of rows of transistor pairs with routing channels between them. The software has freedom to place the cells anywhere within the rows. The placement is optimized so the routing between the cells is minimized. This comprises the core of the circuit. The periphery of the IC contains the input and output pads that will be bonded to the external package pins.

The SOLO 1400 placement and routing system has several stages:

(1) The PLACE routine sets the CMOS transistor pairs into a minimum routing distance arrangement of rows and columns.
(2) The GATE routine fills in the details of the gates.
(3) PINOUT provides on-screen facilities for package type selection and positioning of the bond pads.
(4) The ROUTE utility joins the stages and adjusts the spacing between the rows and columns to accommodate the connecting tracks.
(5) DRAW analyses the physical layout of the chip to produce the mask layer artwork in standard CIF (Caltech Intermediate Format).
(6) The artwork can then be viewed and plotted using the ARTVIEW routine.
(7) PACKAGE checks the package selection and bonds the pins to the pads.

The layout artwork of the complete chip is illustrated in Figure A2.3.

A2.7 Post-layout simulation

It is important that the functioning of the circuit is re-checked after the circuit has been physically laid out, because the interconnecting tracks will add a capacitive loading to the signal lines which increases the delays. This limits the maximum speed at which the circuit can operate, and may cause it to function out of specification. The specified maximum speed for the circuit was a very conservative 5 MHz. All 16 test simulations were repeated, and the maximum simulated clock speed was determined to be 25 MHz. The WAVE utility is used to view the output from the MADS simulator, the output being in the form of signal timing plots. A typical output waveform for one of the functional tests is shown in Figure A2.4.

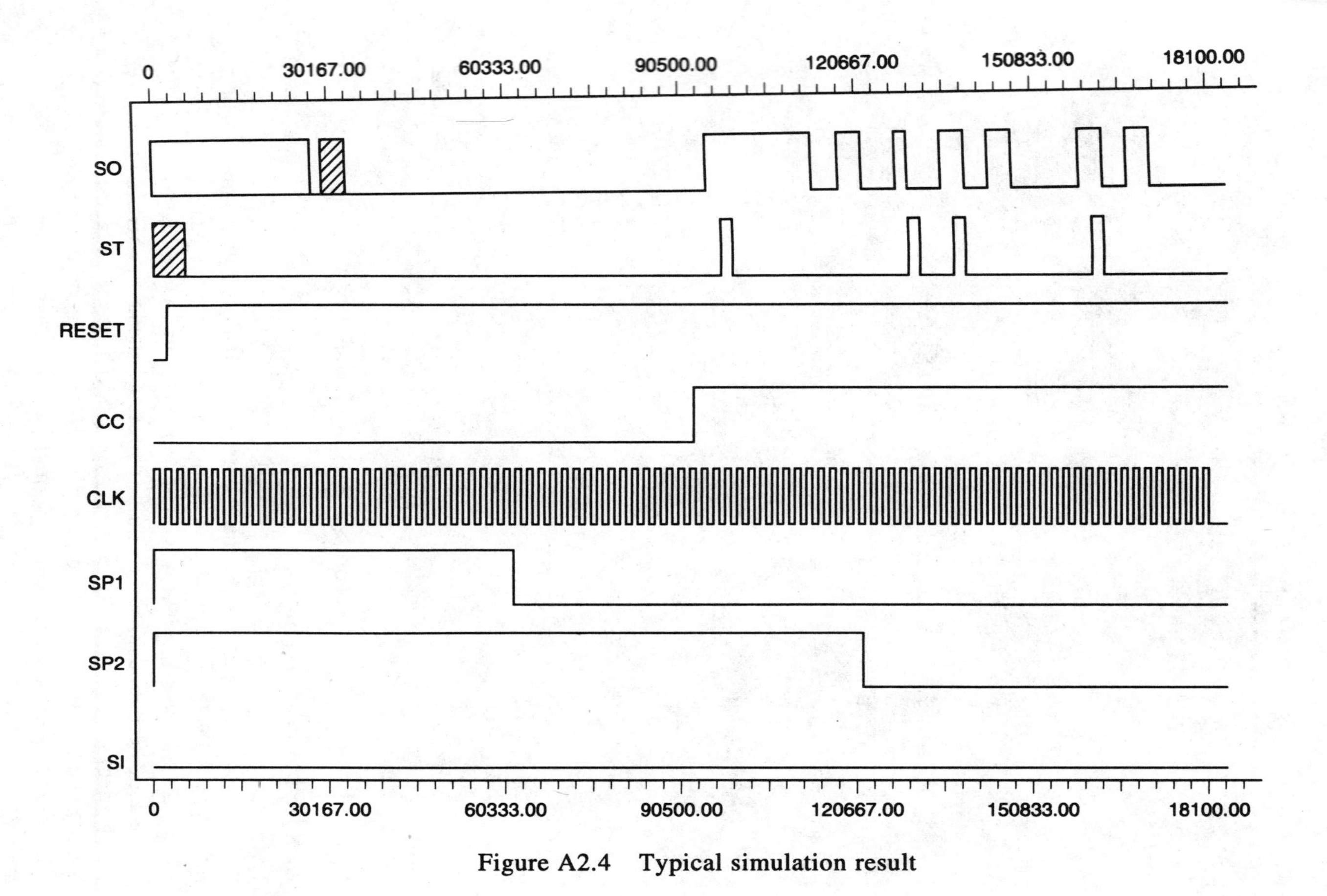

Figure A2.4 Typical simulation result

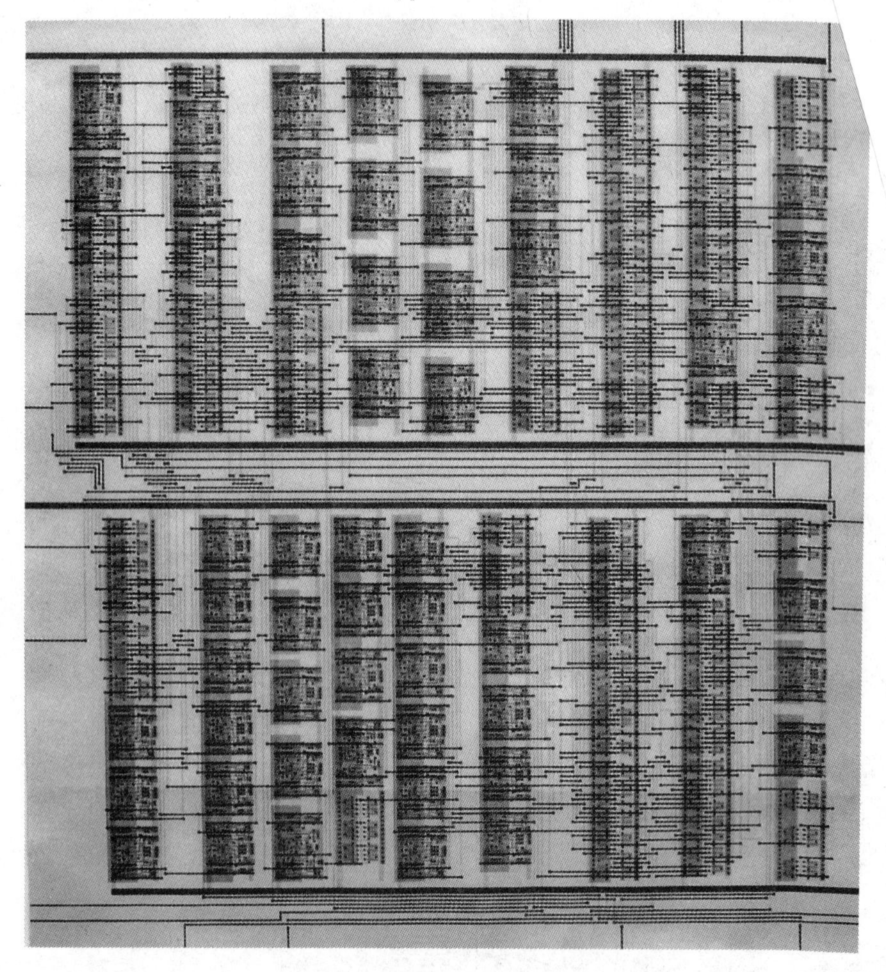

Figure A2.3 Chip layout

A2.8 Design submission

A number of checks have to be made before the circuit design is submitted for fabrication. The circuit has to pass all the checks before the design is accepted by ES2, in order to prevent wasted time and effort in fabricating a design that would be bound to fail. This is performed using the SHIPDES utility which checks that all the design stages have been completed (and in the right order!). It also checks for any errors or warnings that have been flagged by the other utilities. Some warnings (for example, unconnected nets) may be acceptable, for instance the $\overline{Q}$ outputs from flip–flops are often not used, but the designer should check that any warnings are acceptable for the circuit to function correctly.

A2.9 Post-fabrication testing

A number of electrical tests are performed during the fabrication stage by ES2, mainly as a process monitoring and quality assurance exercise. The results of these tests are provided to the customer, including the maximum and minimum acceptable values, the measured values at a number of sampled die sites, and the average and standard deviation from these for each wafer processed.

The designer has to provide a test vector specification with a guaranteed node toggle count in excess of 95 per cent, which can be used to test the circuits before packaging. This allows for only the good circuits to be returned to the customer. The number of packaged circuits depends on the number of designs on the MPW, the process yield and the number passing the testing stage.

A total of nine packaged devices were delivered. These were tested for functionality, along the lines of the functional tests at the simulation stage. All devices passed these tests. The maximum clock rate for the device was typically 24 MHz, which agrees very closely with the post-layout simulated maximum speed of 25 MHz.

A2.10 Conclusion

The exercise described has demonstrated the ease with which engineers with virtually no experience of IC design can realize a semi-custom ASIC. Although this was an academic exercise rather than for commercial production, it covered many of the procedures and considerations that would be required for a successful product.

Index